D1519058

HELPERS AT BIRDS' NESTS

HELPERS AT BIRDS' NESTS

A WORLDWIDE SURVEY OF COOPERATIVE BREEDING AND RELATED BEHAVIOR

BY ALEXANDER F. SKUTCH
ILLUSTRATED BY DANA GARDNER

University of Iowa Press Iowa City

University of Iowa Press, Iowa City 52242

Printed in the United States of America
First edition, 1987

Book design by Patrick Hathcock
Jacket design by Richard Hendel
Typesetting by G & S Typesetters, Austin, Texas
Printing and binding by Malloy Lithographing, Ann Arbor, Michigan

Library of Congress Cataloging-in-Publication Data

Skutch, Alexander Frank, 1904–
Helpers at birds' nests.

Bibliography: p.
Includes index.
1. Birds—Behavior. 2. Familial behavior in animals.
3. Social behavior in animals. I. Title. II. Title: Cooperative
breeding and related behavior.
QL698.3.S55 1986 598.2'56 86-19295
ISBN 0-87745-150-8

To Ian Rowley, pioneer in the quantitative study of cooperative breeding

CONTENTS

PREFACE

Helpful birds serve other birds who are neither their mates nor their dependent young. Most commonly and substantially, they do so by feeding them, but they may warn them of danger, assist in nest construction, incubate eggs or brood nestlings not their own, or preen companions. When the helpers include birds who, after becoming independent, remain with their parents in integrated, stable, friendly groups and attend the parents' subsequent broods, the situation is called cooperative breeding, which is found chiefly in warm countries where migration does not disrupt families. Sometimes the helper feeds young of a different species, an event most frequently reported in the temperate zones, where bird-watchers are more numerous. Occasionally in many species and regularly in a few species, two females or two pairs share a nest and rear their young together, becoming mutual helpers. Helpful birds have been found in the most diverse contexts.

Whether they occur in our tense human world or in nature's wider realm, aggression, strife, and violence seem to fascinate people and have become the subject of many books. The larger and fiercer the predator, the more it engages the attention and claims the funds of conservationists. Some biologists seem to delight in detecting repressed hostility in acts that appear friendly, and they use all their ingenuity to disclose selfish deceit or cheating in ostensibly altruistic behavior. They appear to take perverse satisfaction in exposing the nastiness and harshness of the living world. Harsh it undeniably is, but not so unmitigatedly as it is often painted. Nature has a gentler side that is too frequently overlooked. Ranking high among the more amiable aspects of the natural world is the helpfulness of birds. For several decades, ornithologists have been studying their cooperative societies with increasing thoroughness. The results of these investigations have mostly been published in scientific journals not readily accessible to, or easily readable by, a wider public. To make helpful birds more generally known, as a possible antidote to all the publicity that nature's uglier side receives, is one aim of this book.

A few of the birds in this book will, I hope, be familiar to readers wherever they live. Unless they have traveled far or read much about birds, most, I fear, will not. To overcome this difficulty, I have included a brief account of the appearance and habits of the species that chiefly claim our attention, trying to place their helpful activity in its proper context, without which it means little. As far as possible, we have included illustrations of the principal actors. Except oceanic birds, precocial birds, and diurnal raptors, of which I have little to report about any single family, I have treated the helpful birds by families, briefly characterizing each in an introductory paragraph. When a species is first mentioned, its English name is followed by its scientific name, which can also be found in the index. References at the end of each chapter are cited in full in the bibliography.

I am grateful to all those colleagues who, unsolicited over the years or more recently at my request, have sent me offprints of their papers on cooperative breeding and related phenomena. I believe that I have covered rather thoroughly the literature on cooperative breeding, and I apologize to any author whose important observations I have omitted because, since I write at a great distance from a well-stocked ornithological library, they have been inaccessible to me. I am especially indebted to the Western Foundation of Vertebrate Zoology in Los Angeles, California, for photocopies from books and journals that have helped greatly to make this book complete.

HELPERS AT BIRDS' NESTS

1. INTRODUCTION

Half a century ago, I studied birds on a farm that stretched from the flatlands at the edge of the Motagua Valley back into the foothills of the Sierra de Merendón, in what was then disputed territory between Guatemala and Honduras. Among the most conspicuous birds in the banana plantation and light surrounding woods were big, noisy Brown Jays. Although I could not distinguish individuals by their plumage, I could do so by their bills, which were black, yellow, or these colors combined in the most diverse patterns. At first largely yellow, the bills blackened as the birds grew older.

While I walked along a ridge in the foothills on a sunny morning late in April, I noticed a bulky nest of sticks in the top of a small tree growing amid scrubby vegetation on the steep slope below me. In it sat a Brown Jay whose bill was more yellow than black. While I stood looking over the tree, considering the feasibility of climbing it to see what the nest contained, a bird with an almost wholly yellow bill flew up, protested my presence rather mildly for a Brown Jay, then alighted on the nest's rim and fed the incubating female. A few minutes later, this bird returned with more food for the jay on the nest. Doubting that a jay with so yellow a bill could be the incubating bird's mate, I sat in a grassy opening near the crest of the ridge, slightly higher than the nest, and continued to watch. After nearly an hour, the incubating bird began to repeat cries which drew a black-billed jay, apparently her mate, who passed her something that he brought in his swollen throat. So at least two full-grown jays were attending this incubating female. Profoundly stirred by this observation, which immediately suggested to my mind a situation rare if not new in the annals of ornithology, I resolved to intensify my study of these jays.

The Brown Jays' bulky nests, attended by birds who were far from silent, were not difficult to find in the banana plantation or the surrounding second-growth woods. At each of four that I watched while they held young, from three to seven grown birds fed and guarded them. Whatever the number of attendants, the brood never consisted of more than three nestlings, apparently the progeny of a single mated pair, who were assisted by one to five individuals with bills largely yellow. These auxiliaries appeared to be chiefly yearlings who, instead of breeding, helped pairs of older birds. Later, in Costa Rica, I watched other Brown Jay nests, each of which had more than two attendants.

The following year, in the high mountains of Guatemala, I discovered two other species with helpers. One was the big Banded-backed Wren, which builds bulky nests with a side entrance amid lowland rain forests as well as on high, frosty mountains. The other was the tiny Bushtit, in each of whose charming pensile pouches four nestlings were fed by three to five grown birds.

In a paper called "Helpers at the Nest," published in 1935, I told about these three species of birds with unusual nesting habits, among the first reported examples of a system of reproduction that has since been designated cooperative breeding. With the notable exceptions of David E. Davis' studies of anis, Ian Rowley's first paper on Superb Blue Wrens, and Robert Norris' monograph on nuthatches, little was added to our understanding of cooperative breeding during the quarter century that intervened between this date and the publication of my second paper on the subject, "Helpers among Birds," in 1961. Subsequently, interest in this phenomenon has increased greatly, and many more examples of this system of reproduction, in the most diverse families of birds, have been painstakingly investigated by ornithologists, chiefly in the tropics and subtropics of America, Africa, Asia, and Australia, less often in the temperate zones. So widespread is cooperative breeding that, as more people become interested in tropical birds, fresh examples are reported almost yearly, and its importance as a method of population adjustment, no less than for evolutionary theory, has incited much learned discussion. As evidence of the recency of interest in this subject, John Terres' *Encyclopedia of North American Birds,* published in 1980, contains a long entry on "Helpers among Birds," but the British Ornithologists' Union's comprehensive *New Dictionary of Birds,* published in 1964, has no article on either helpers or cooperative breeding.

In my pioneer paper on the subject, I gave the name "helpers" to birds that aided mated pairs. Although birds that participate in cooperative breeding associations are, biologically and ecologically, the most important category of helpers and will receive the major share of our attention in the following pages, they are by no means the only ones that merit our notice. In my second paper, I wrote: "A 'helper' is a bird which assists in the nesting of an individual other than its mate, or feeds or otherwise attends a bird of whatever age which is neither its mate nor its dependent offspring. Helpers may be of almost any age; they may be breeding or nonbreeding individuals; they may aid other birds of the most diverse relationships to themselves, including those of distinct species; and they may assist in various ways." Accordingly, to understand helpfulness among birds, we must give attention to the following points: the status or condition of the helper, whether young or old, a parent or a nonbreeder; its relationship to the bird or birds whom it assists; and the activities in which it engages.

First, as to the helper's status. From hatching until it leaves the nest, a bird is a "nestling." "Fledgling" or "chick" designates a bird that has left the nest but is still dependent upon parental care. In the months immediately following the attainment of independence, a bird is a "juvenile." In the breeding season immediately following that in which it hatched, when it is approximately one year old, it becomes a "yearling." "Nonbreeding adult" designates all full-grown birds not directly engaged in reproduction, whether they are or are not physiologically able to breed—a point often difficult to determine. "Breeding adults" are mature individuals engaged in the production of their own descendants. They may be about one year old, and therefore "yearlings," but to avoid confusion we shall reserve this designation for individuals who do not breed, those who in many cooperative nesting associations are the most numerous class of helpers. Among marine birds and many others that do not attempt to reproduce until two years of age or often much more, nonbreeding adults comprise a large proportion of the population.

Helpers fall into two great divisions with reference to their relationship to the birds they assist. They include "intraspecific helpers," who aid others of their own kind, and "interspecific helpers," who assist individuals of different species. With these distinctions, we may attempt to classify helpers as follows.

I. In reproductive activities
    A. Intraspecific helpers
        1. Nestling helpers
        2. Juvenile helpers
        3. Yearling helpers
        4. Nonbreeding adult helpers
        5. Breeding helpers
            a. Unilateral helpers
            b. Mutual helpers
    B. Interspecific helpers
        The subdivisions under this heading are the same as those under A. Except in aviaries, where juvenile helpers are frequent, most known examples belong to subdivisions 4 and 5.
II. In nonreproductive activities
    A. Intraspecific helpers
    B. Interspecific helpers
        The examples under both A and B that are at present available are so few that no further subdivision seems advisable.

When parents and their assistants are integrated into coherent, harmonious, fairly stable social groups that persist throughout the year, the situation may be designated advanced cooperative breeding. Not all nesting pairs of species properly called cooperative breeders have helpers; those nesting for the first time and those that have nested unsuccessfully or have lost all their fledged young are most likely to be without assistants, who are usually, but by no means always, attached to their own parents; but in all cooperative breeders a substantial proportion of nesting pairs, commonly a third or more, have at least one helper. Unilateral helpers, who assist others with no immediate reward (although their association with their beneficiaries may aid them in various ways), are the most frequent kind in both intraspecific and interspecific contexts; as juveniles or yearlings, they are the mainstay of many cooperative breeding associations.

Although some authors have used

"communal breeding" instead of, or to be synonymous with, "cooperative breeding," we shall stay closer to the meaning of "communal" if we restrict this designation to groups in which two or more females lay their eggs in the same territory, whether in one or several nests, and assist each other to rear their broods. (Most advanced cooperative breeders are families; we do not call a human family a community.) Communal nesters may be advanced cooperative breeders, but most advanced cooperative breeders do not nest communally.

Some writers have avoided the term "helper" on the grounds that it is anthropomorphic and, since its introduction into the ornithological literature in 1935, has been applied to cases of casual or interspecific assistance. Veronica Parry, writing of the social system of Laughing Kookaburras, preferred the term "auxiliary." Perhaps we could agree to restrict this designation to nonbreeding birds who assist in cooperative breeding associations. "Helper" would continue to be the more general term, inclusive of auxiliaries among cooperative breeders. All auxiliaries are helpers, but not all helpers are auxiliaries. In many situations, we may use the two words interchangeably to avoid monotonous repetition.

If we give the term "helper" the widest possible inclusiveness, we must recognize that a large proportion of all birds are helpers, for they aid one another, if in no other way, by promoting mutual security. Among the explanations of the mixed flocks of few or many species that wander through both temperate-zone and tropical woodlands, the most convincing is that members of such flocks feel, and are, safer than they would be if they foraged alone. At first sight, this appears improbable, because such companies of active, chattering birds are so conspicuous. However, it appears that many watchful eyes, many voices quick to sound the alarm, more than compensate for conspicuousness, and predators find it easier to surprise solitary birds than to capture members of such flocks.

Among open-country birds, the warning cry of one alerts others within hearing to the approach of danger; they stop singing and seek cover as a hawk flies overhead. Not only birds of diverse species but also mammals, such as deer, learn to recognize these alarm signals and to take measures for their safety. Mobbing, which occurs when a variety of birds flit around a predator, often a snake or a somnolent owl, protesting its presence in many different tones, promotes the safety of the avian community because it is always advantageous to know just where the enemy is—although rarely a too-reckless mobber is caught by the object of its displeasure. Mobbing also has educational value: by participating in these commotions, young birds learn to recognize potential enemies. Birds that nest in colonies, and even those of different species whose nests are close together, may unite in threatening or attacking undesirable intruders. More rarely, nesting birds try to lure marauding animals from their nests by simultaneous distraction displays. These ways of helping are impersonal, benefiting an avian community rather than being directed by one individual to another individual or pair of birds, and we need not consider them further.

Among cooperative breeders, helpers of all ages past the fledgling stage commonly join the parents in defending the group's territory against neighboring groups. Likewise, they try to protect nests and fledglings from predators, sometimes with success. Of the more intimate ways of helping that will chiefly claim our attention in the following chapters, by far the most frequent is feeding. Because the perpetuation of most avian species depends upon direct feeding of the young, the impulse to feed has become strong and enduring in birds. It often appears in juveniles and even nestlings, and it may persist in individuals no longer able to breed. It frequently oversteps the limits of species, giving rise to interspecific helping. Although primarily directed to the nourishment of young, feeding has become important in courtship and maintenance of the pair bond in many species; usually the male offers food to his mate, but sometimes she returns the courtesy. By feeding the incubating female, mated males and often, too, their helpers enable her to sit more constantly or even continuously. This is the last manifestation of parental behavior to persist in parasitic birds. Among altricial, semialtricial, and subprecocial birds, feeding the young is nearly always the chief contribution of helpers.

Although precocial birds do not feed their young directly, they brood, guide, and lead them to food and may become helpers by adopting chicks of their own or different species, as has frequently been recorded of waterfowl, gallinaceous birds, and shorebirds. This may save the lives of many chicks and ducklings who have lost or become separated from their own parents.

Among less frequent modes of helpfulness are nest building, incubation, brooding, and nest sanitation—removing nestlings' droppings. Although juveniles and even nestlings sometimes make gestures of nest building, the construction of anything more complex than hollows scratched into sand or loose soil demands more skill than is required to pass food from one bill to another, and their efforts are often ineffectual. When nests are maintained throughout the year as family dormitories, somewhat older helpers may contribute importantly to their construction or repair. Cooperation in building is frequent among mutual helpers, including those that construct large nests with several or many chambers—Monk Parakeets, Palmchats, and Sociable Weavers—and communal nesters such as anis. The large mounds in which scrub fowl, or megapodes, deposit their eggs, to be incubated by the heat of a great mass of decaying vegetation, are kicked together by several pairs of birds, who thereby qualify as mutual helpers. Among communal nesters, participants regularly share incubation, but in other species helpers incubate much less frequently than they feed the young. As a rule, individuals that incubate also brood the nestlings.

Among the modes of helping not directly connected with reproduction we may include allopreening or the grooming of one individual by another. Mates who preen one another and parents who preen their young are not helpers as we define them, but young who groom their parents or siblings are. Likewise, highly social birds who preen flock members other than their mates deserve to be included in this category. In captivity and more rarely when free, birds sometimes preen individuals of different species, thereby swelling the list of interspecific helpers. A particularly endearing way of helping is "putting the fledglings to bed" or guiding and encouraging them to enter the nests where they sleep, often difficult for weakly flying young birds to reach.

In chapter 49 we shall meet a Northern Cardinal who fed seven goldfish in a pond. His activity so greatly resembled that of a bird who feeds the young of another avian species that I did not hesitate to include him among interspecific helpers, although he hardly conforms to our definition of a helper as one who attends another bird. If we admit the eccentric cardinal among the helpers, why must we stop short with the animal kingdom—why not broaden our category to include all those birds that benefit plants, the nectar drinkers who pollinate flowers, the frugivorous birds who disseminate seeds, perhaps even insectivorous species that relieve plants of the creatures that devour their living tissues? But probably this would be stretching the category too far. All these birds are well paid for their unintentional services with the nectar, fruits, or insects that sustain them. The helpers that we shall meet in this book are never paid directly for their aid, although their indirect benefits may be considerable—as is sometimes true when we perform altruistic acts. Under the most liberal interpretation, all members of a cooperating group, the parents no less than their auxiliaries, are mutual helpers, for as we shall see they contribute to each other's welfare in diverse ways—which is as it should be, for true cooperation is a reciprocal relationship.

REFERENCES: D. E. Davis 1940a, 1942; Frith 1956; Norris 1958; Parry 1973; Riney 1951; Rowley 1957; Skutch 1935, 1961, 1976; Terres 1980; Thomson 1964.

2.

OCEANIC BIRDS

Although, after attaining adult size and independence, many marine birds delay reproduction for several years—a situation often associated with cooperative breeding in terrestrial birds—their way of life is incompatible with the persistence of family bonds that leads to cooperation. They nest in crowded colonies on small islands or forbidding seaside cliffs which provide no food, and after breeding they scatter over wide expanses of the open ocean, where it would be difficult for families to remain intact. When pairs persist from year to year, they appear to be reunited at the nest site after months of separation on the high seas. Yet, despite a way of living that fails to promote cooperative breeding, we find among them instances of helpfulness worthy of attention.

About the time the last chicks are hatching in a colony of Adelie Penguins (*Pygoscelis adeliae*) at the icy edge of the Antarctic continent, nonbreeders two or three years of age and older birds whose nests have failed return from the sea for what is called the reoccupation period. They have been regarded as a menace to the young chicks, for while struggling among themselves for possession of abandoned nests they injure or kill some of them. Recent observations by Yasuomi Tamiya and Masahiro Aoyanagi, however, suggest that they may do more good than harm. South Polar Skuas (*Catharacta maccormicki*), nesting around the edges of the penguin colony, prey heavily upon the penguins' eggs and chicks, which must be constantly guarded lest these rapacious birds carry them off to feed their own young. Occupying empty nests at the margin of the colony or in its midst, the nonbreeding penguins who have returned from the sea repel the skuas, sometimes attacking them violently. They incubate eggs and brood young chicks temporarily left uncovered by their parents, protecting them from the skuas and keeping them warm until a parent returns to reclaim them and drives the helper away.

Soon after the young penguins hatch, they gather into crèches, where on cold days they huddle together for warmth but in milder weather spread apart, while their parents seek food for them in the sea. The nonbreeders guard these crèches and with pecks drive straying chicks back to them, where alone they are safe from the skuas. After the chicks are somewhat older, a parent returning from the water calls its own offspring from the crèche and feeds it 10 to 20 yards (9 to 18 meters) away, to avoid interference by other chicks or to give it exercise by walking. After regurgitating food to the youngster, the parent marches off, leaving it perilously exposed. Now nonbreeding guardians, who have followed the chick to the spot where it was fed, push and peck it back to safety in the midst of its compeers. Without the often unruly nonbreeders, losses of young penguins might be much greater.

Downy chicks of a number of birds that nest on the ground or in shallow water in colonies, including certain penguins, pelicans, and flamingos, gather into crèches while their parents seek food for them. Earlier observers, concluding that these massed young birds were fed indiscriminately by parents unable to recognize their own amid the crowd, were impressed by what appeared to be an admirable example of cooperation. A little reflection will convince us that this would lead to great injustices—without a degree of regimentation that would be difficult to im-

Adelie Penguins

pose even upon young children in the open, the more vigorous chicks would claim most of the food and the smaller or more timid would go hungry. Actually, by voice and appearance, parents and chicks recognize each other and, with rare exceptions, each adult feeds his or her own young. Individual attention to its own offspring by each parent insured that even a young Greater Flamingo (*Phoenicopterus ruber*) with a broken wing received its meal, although in a general melee to get food it would probably have been pushed aside.

On bleak South Georgia Island, where King Penguins (*Aptenodytes patagonica*) breed, Bernard Stonehouse found that a young chick's chances of survival decrease rapidly after five minutes of exposure on a cold, windy day. Such a chick, temporarily neglected by its parents, is often picked up by an unemployed adult, who rolls it onto its feet and broods it. If several such adults are present, they may pass the chick from one to another, keeping it alive until one of the parents reclaims it. Large chicks may nurse tiny chicks with similar brevity. Paradoxically, the tiny chick's survival depends upon the inconstancy of its benefactors: a temporary guardian who held a foundling for several hours would reduce the probability of its retrieval by one of its parents and find itself in charge of a chick that it could not rear. Sometimes a third adult joins the parents in incubating the King Penguin's single egg and feeding the chick. With three attendants the chick gains weight faster than those nourished by the two parents alone. Unemployed Emperor Penguins (*A. fosteri*) also pick up and place on their feet, under the brood pouch at the base of the abdomen, chicks that have been left exposed, often struggling among themselves for possession of foundlings. By a succession of such fosterers, a chick may be saved from freezing on the Antarctic ice field.

On Midway Atoll in the mid Pacific, parent Laysan Albatrosses (*Diomedea immutabilis*) deserted their nestling after Eugene Lefebvre attached radio transmitters to both of them. Nevertheless, the chick continued to grow and remained normally active. Investigation of this puzzling situation revealed that the young albatross was attended during the next three months by no fewer than six helpers. One merely sat with it for five weeks, preening it repeatedly. Five other adults fed the chick every two or three days. One brought food for two weeks, then ceased to attend it. After it grew bigger, the young albatross was fed daily, six times on one day by several adults. By the combined efforts of a number of volunteer attendants, the orphaned albatross was nourished so well that after the three months it moved toward the beach, apparently ready to fly over the ocean.

A chick that, by precociously engaging in adult activities, lightens the burden of parental care may without exaggeration be called a helper. Even when newly hatched, chicks of the Black-footed Albatross (*D. nigripes*) of Midway Atoll kick sand out of their nest hollows on the open beach, an activity that may save their lives when storms blow loose sand into them. If the gale continues until they become exhausted, they may be buried and die. When they grow too large to be comfortably brooded, the young albatrosses may make a new nest depression beside the original nest where a parent is sitting. Still later, after the parents cease to guard it continuously, a chick may dig one or more new nests up to 100 feet (30 meters) from its natal spot, often in a shadier place. It lies in such self-made nests only by day and returns to the original nest to pass the night.

Similarly, from an early age, nestlings of the Short-tailed Shearwater (*Puffinus tenuirostris*) collect grasses and debris from the walls of their chamber at the end of a burrow, then tuck them around themselves to form a neat cup in which they sit. Until they leave the nest, they continue this behavior, with the result that they accumulate a considerable mass of material when they venture into the entrance tunnel and carry pieces back to the nest.

When a five-day-old chick of the Red-billed Tropicbird (*Phaethon aethereus*) rolled down a cliff into a nest containing a fifteen-day-old chick, the two were reared together. It was possible, but unlikely, that the foundling was fed by its own parents or that both pairs of parents fed both chicks. In each of two nests, a chick of the Yellow-billed Tropicbird (*P. lepturus*) was hatched and reared by Red-billed parents, who had stolen the nest site from the Yellow-bills.

Like albatrosses, young cormorants engage in nest building at an early age. Alarmed chicks of the Shag (*Phalacrocorax aristotelis*) often abandon their nest and station themselves in a new site several yards away, where a new nest grows up, with materials brought not only by returning parents but also by the chicks themselves. When a month old, Shag chicks briefly preen their parents and each other. After they leave the nest, they may also preen other broods. Parents preen alien juveniles who assume submissive postures. Nestlings of the European Cormorant (*P. carbo*) feed one another and help work loose pieces into their nest.

A Magnificent Frigatebird (*Fregata magnificens*) with only one wing was fed by other members of its breeding colony. Similarly, an adult Brown Booby (*Sula leucogaster*) who had lost a wing was kept alive with food that its neighbors in a breeding colony supplied. In a colony of American White Pelicans (*Pelecanus erythrorhynchos*) an old, blind individual was found alive. Since it could not feed itself, it must have been supported by others. These three handicapped adults must have begged compellingly from indulgent neighbors.

South Polar Skuas with eggs or young of their own sometimes accept and rear chicks that wander from neighboring nests, from which the younger of two chicks is often driven by its older sibling. Such adoption usually results in the desertion of the fosterers' eggs or the death of their own offspring, for they are usually unable to rear two chicks.

Perhaps it will not be inappropriate to include here gulls that breed beside inland waters as well as those that nest on marine islands. Gulls have long been known to be monogamous, but unconventional breeding associations have recently been discovered among them. In colonies of Western (*Larus occidentalis*), Ring-billed (*L. delawarensis*), California (*L. californicus*), and Herring (*L. argentatus*) gulls, a small proportion (not over 1 or 2 percent) of the nests are attended by two females, who behave like a mated pair, with no male attached to them. Nests of these female-female pairs often contain five or six eggs instead of the gulls' usual clutch of two or three, and since many of these eggs hatch, they must have been fertilized by males that were not the females' mates. In one of the rare cases of polygyny reported for gulls, three females and one male Ring-bill incubated in a nest in which at least two females had apparently laid. These deviations from the gulls' usual breeding system seem to result from a small excess of females over males. They remind us of the communal nesting of Magpie-Geese (*Anseranas semipalmata*) or of anis, but we still know too little about them to assess their importance.

In a ringed population of Arctic Terns (*Sterna paradisaea*) on the Farne Islands in the North Sea, most of the birds first bred when three years old. Unable to nest because of a shortage of females, some of the younger males helped feed the offspring of established pairs. In

Common Murres

times of scarcity when mortality among chicks is high, terns of various species feed those of other parents.

Murres, called guillemots in Britain, nest in crowded colonies on ledges of high seaside cliffs or on small, remote islands. Each female lays a single egg whose pyriform shape causes it to roll in a small circle, thereby reducing the probability that it will fall off the ledge or into a crevice from which it cannot be retrieved—a major source of egg loss. When five or six or more days old, chicks of both the Common Murre (*Uria aalge*) and the Thick-billed Murre (*U. lomvia*) often take shelter beneath the wings of adults other than their parents, sometimes as many as three or four under a single bird. Bereaved adults and even those with young brood and feed chicks not their own, although the true parents sometimes repel neighbors' attentions to their offspring. Before they can fly, young murres flutter down to the sea and swim away accompanied by an adult of either sex. Although it is sometimes asserted that the chick's escort is a parent, Leslie Tuck's observations of banded individuals of both species of murres revealed that it is much more often some other adult. How long the volunteer attendant guides and feeds the young murre on the open ocean we do not know.

References: L. H. Brown 1958; Conover et al. 1979; Cullen 1957; Lefebvre 1977; Murphy 1936; Perry 1946; Rice and Kenyon 1962; Ryder and Somppi 1979; Shugart 1980; Skutch 1961; B. K. Snow 1963; Stonehouse 1953, 1960, 1962; Tamiya and Aoyanagi 1982; Tuck 1960; Warham 1960; E. C. Young 1963.

3.

PRECOCIAL BIRDS

Precocial birds include all those whose young leave the nest within a day or two of hatching and are guided by their parents while they pick up their own food, instead of being fed directly by the adults, as in altricial, semialtricial, and subprecocial species. Because of the less elaborate parental care of precocial birds, such helpers as occur among them assist in less diverse ways than altricial helpers do. Precocials most frequently earn the designation of helpers by permitting strayed or orphaned young of other parents to join their own families and benefit from their guidance and protection, including probably brooding, a point difficult to ascertain in free birds.

The polygynous Greater Rhea (*Rhea americana*) incubates his large clutch of eggs and cares for the chicks with no help from the females of his harem. As he leads his young flock over the grassy Argentine pampas, brooding it and protecting it by attacking animals that menace it or luring them away by impressive distraction displays, he often collects lost or orphaned chicks, until he is followed by a group ranging in age from the newly hatched to gangling rheas two months old. Unlike rheas, both sexes of the Ostrich (*Struthio camelus*) incubate the eggs and attend the young, but one of the chicks' guardians may be a stepparent, with a flock that contains young from different nests. To lure predators from their dependents, the huge Ostriches simulate injury much as smaller birds do. Ostriches practice communal nesting, at least in rudimentary form. The dominant female, who incubates by day, while the single male does so through the night, sometimes permits subordinate females who have contributed eggs to the combined clutch to take turns on the nest.

Among gallinaceous birds, hens of the Ruffed Grouse (*Bonasa umbellus*) occasionally lead more chicks than they hatch—up to fifteen in a family of mixed parentage—or young of different ages, proof that they have accepted offspring of other parents. A grouse hen who lost her eggs just before they were due to hatch somehow acquired and mothered four chicks. In Alberta, Canada, 11 young Spruce Grouse (*Dendragapus canadensis*), 4.1 percent of 266 marked chicks, joined parents not their own. Among them were 8 orphans. All these changelings survived well until autumn.

Among shorebirds, male Stilt Sandpipers (*Calidris himantopus*) and Baird's Sandpipers (*C. bairdii*) sometimes attend young not their own, who without their foster parents would probably succumb. American Avocets (*Recurvirostra americana*) and Black-necked Stilts (*Himantopus mexicanus*) also lead mixed families, with young of different ages, or a number larger than their sets of eggs. Some adults try to "kidnap" chicks from such enlarged broods. Parent European Avocets (*Recurvirostra avosetta*) brood recently hatched chicks of their neighbors. Both sexes of the Snowy, or Kentish, Plover (*Charadrius alexandrinus*) take an interest in chicks of other members of their own and of related species and may even brood them.

Among waterfowl, adoption is not rare. In Wisconsin, between one-third and one-half of all pairs of Canada Geese (*Branta canadensis*) accepted foreign goslings from a few days to eight weeks old, but usually under two weeks of age. A wild male Canada Goose accompanied and guarded a brood of thirteen (domestic?) ducklings, along with the parent duck, "who seemed to welcome the gander." He followed the family in the rear and stayed with it all day. Albert Hochbaum, an authority on waterfowl, believed that in all species of ducks a parentless brood may occasionally join another in its entirety, and this process may be repeated until a large aggregation is formed. He saw a White-winged Scoter (*Melanitta fusca*) with a flotilla of eighty-four ducklings, all under two weeks old. Lesser Scaup (*Aythya affinis*) often lead twenty or more ducklings, sometimes with two or more female ducks attending the combined families.

When Shelducks (*Tadorna tadorna*) desert their broods to migrate to the distant shallow seas where they molt, the resulting crèches may become spectacularly large, with over one hundred ducklings attended by a few faithful adults. To the crèches of Common Eiders (*Somateria mollissima*) two functions have been ascribed. The female incubates so constantly, leaving her nest only every second or third day for an interval no longer than ten or fifteen minutes, during which she drinks but does not eat, that she becomes emaciated. Nearly half of the annual mortality of females occurs in June and July, just after the incubation period. Leaving their ducklings in the care of helpers, including nonbreeders or females who lost their eggs early, the emaciated mothers can join the main feeding flock and replace the body reserves consumed while they incubated. On this view, the chief beneficiaries of the crèching system are the parents rather than the ducklings.

Others have emphasized the value of crèches in protecting the ducklings from ravenous gulls. In the estuary of the Saint Lawrence River in Canada, J. Munro and J. Bédard found crèches with up to sixty ducklings attended by one to five adult female Common Eiders, but most commonly they contained seven to nine ducklings with two females. These assemblages resulted from chance encounters of mothers leading their broods from nests on wooded islands to rearing areas, which might be as much as 9 miles (14.5 kilometers) away on the opposite shore of the broad estuary, or from the clumping of families menaced by Herring Gulls (*Larus argentatus*). Often the families separated after a brief union, perhaps with some exchange of ducklings; but if they remained together until the ducklings were older and recognized one another and their guardians individually, the crèches became stable groups that often resisted the intrusion of aliens.

Gulls attack ducklings singly, in loose pairs, or in larger groups. Crèches reduce the success of gulls working alone, but a mass attack may annihilate them. The larger the crèche, the more vulnerable to concerted predation it appears to be. However, taking the types of predation together, crèches do reduce losses of ducklings to gulls. The guardians of these crèches are broody females who have lost their eggs. Sometimes one of them so dominates her associates that they abandon the crèche to join another or to forage with unattached eiders. These less aggressive eiders who desert crèches are likely to be individuals whose vitality has fallen while incubating. The two interpretations of the value of crèching by eiders are not mutually exclusive. According to the resources of the habitat and the intensity of predation by gulls, either the parents or the ducklings may be the chief beneficiaries.

The strong tendency of Common Eiders to defend ducklings was impressively demonstrated when five females came to the rescue of twelve downy young of the Northern Shoveler (*Anas clypeata*), menaced by a Herring Gull while their mother led them across open water. The eiders formed a close circle around the shoveler and her brood and rose out of the water each time the gull attacked, continuing this defense until the would-be predator desisted and flew away.

Widespread on the savannas and wet grasslands over the length and breadth of South America, the Southern Lapwing (*Vanellus chilensis*) is a handsome plover with a high, thin crest. Long ago, W. H. Hudson described the elaborate displays that occur when pairs are visited by a third individual. He knew several instances of a male with two females, who laid their eggs in the same nest and took turns incubating them, and likewise of two males with one female. More recently, Jeffrey and Beverly Walters found cases of cooperative breeding by Southern Lapwings in Venezuela. Most nests were attended by unaided pairs, but at four of twenty that were carefully watched auxiliaries were present. A group of one male and two females attended successively two clutches of five and six eggs instead of the usual two to four. All three of the attendants participated in all parental activities, defended the territory, and—two or three together in all possible combinations—engaged in the usual coordinated displays. Another group of three adults of undetermined sex led three chicks who picked up their own food. The third group consisted of three individuals, also of unknown sex, whose nest with three eggs was soon destroyed. Group four was composed of a mated pair and two other birds, one of whom helped incubate the infertile eggs, despite opposition by the primary adults. Another member of this group, a juvenile who was possibly the offspring of the mated pair, joined the adults in attacking predators and defending the territory but took little interest in the nest.

Another plover that occasionally engages in cooperative breeding is the Little Ringed (*Charadrius dubius*), which nests from England to Japan. A nonbreeding bird, at least a year old, frequently assists a breeding pair, helping defend their territory, taking turns at incubation, caring for the chicks, and protecting them from enemies.

It may not be out of place to consider here three species which, although not strictly precocial, belong to predominantly precocial families that we have already met in this chapter. The first is the Common Snipe (*Gallinago gallinago*), which lives in wet, grassy places, bogs, and marshes over much of the world. Although, like other shorebirds, a day-old snipe chick pecks at small objects on the ground, its short bill is ill adapted for gathering food from the mucky soil, and for about the first week of its life it is almost wholly dependent on its parent, who feeds it directly from the bill in the manner of subprecocial birds. If the young snipe finds a complaisant adult, it may solicit, and receive, bill feeding much longer. A three-week-old chick captured in the wild and placed with a two-year-old hand-raised adult pursued the older bird, squeaking and trying to snatch food from its bill, until the harassed adult began to feed the youngster liberally, thereby joining the category of helpers.

Like the snipe, the chick of the European Woodcock (*Scolopax rusticola*) has a bill too short to feed itself effi-

Magpie-Geese

ciently before it is twelve or thirteen days old. In captivity, a twenty-six-day-old female foster parent fed earthworms to newly hatched chicks. Two males did the same when ten months old.

The large black-and-white Magpie-Goose (*Anseranas semipalmata*) of northern Australian swamps breeds in trios of a male and two females more often than in simple pairs. The two females of a polygamous male lay a total of seven to twelve eggs in a nest which both build, apparently with his help. Then all three take turns incubating the eggs for twenty-four or twenty-five days, and all care for the goslings, at first feeding them from the bill, as no other member of the duck family is known to do. These aberrant geese have much in common with communally nesting anis, except that in the latter each of the females who share a nest has her separate mate. In other species of waterfowl, such as the Wood Duck (*Aix sponsa*) and the Mallard (*Anas platyrhynchos*), two females occasionally lay their eggs in the same nest and share incubation, but only in the Magpie-Goose is this usual.

References: Bellrose 1943; Bruning 1975; Duebbert 1968; Frith and Davies 1961; R. B. Hamilton 1975; Hochbaum 1960; Hudson 1920; Jehl 1973; Keppie 1977; Marcström and Sundgren 1977; Maxson 1978; Messenger 1949; Milne 1969; Munro and Bédard 1977a, 1977b; Sauer and Sauer 1966; Selous 1927; Tuck 1972; Vaughan 1980; Walters 1959; Walters and Walters 1980; Weatherhead 1979; Zicus 1981.

4.

DIURNAL RAPTORS

About 260 species of hawks, eagles, kites, falcons, and their allies are distributed over all the continents except Antarctica and on many remote islands. Among the latter are those of the Galápagos Archipelago, in the Pacific Ocean 600 miles (965 kilometers) west of Ecuador, where the Galápagos Hawk (*Buteo galapagoensis*) is endemic. It lives in single pairs or, more frequently, in polyandrous groups, with two or three, rarely four, males mated to a single female. One nesting association consisted of five males and two females. The members of these breeding groups tend to remain together on the same territories from year to year. Although in some years adult males appear to be more numerous than adult females, the difference in the numbers of the two sexes in the total population can hardly account for the fact that more than twice as many males as females enter the breeding population, leaving many mature females unmated. The available habitat appears to be too small to accommodate all potential breeders. Male Galápagos Hawks solve the difficulty by joining in polyandrous groups with usually a single female; but the females, as in many raptors the larger, more aggressive sex, are apparently too intolerant of other females to share a nest or a territory with them often.

All the males in a breeding group help defend their territory, incubate, and pro-

Galápagos Hawk

vide food for the female and the single set of young. Each may procreate one or more of the one to three nestlings, and each feeds the other's offspring as well as its own, so that these polyandrous males conform to our definition of helpers. Nests of monogamous pairs yielded, on the average, only one fledgling, whereas those attended by polyandrous groups produced two. However, monogamous males did better than polyandrous males, with 1 young per male for the former instead of only 0.6 young for the latter. According to evolutionary theory, the male hawks should choose the reproductive system that gives them the greatest number of living progeny. Why do they settle for less? Apparently, they share a territory and a female with other males because it increases their probability of survival. Three to five months after they fledge, young Galápagos Hawks are driven from the parental territory; they do not acquire territories and breed until at least two or three years later. The hawks without established territories suffer much higher mortality than those in territorial groups. The sooner an individual can acquire a territory, usually by joining a polyandrous association, the greater his prospects of survival. By prolonging his years as a breeder, he compensates for the smaller number of his own offspring from each nest.

The breeding system exemplified by the Galápagos Hawk and the Tasmanian Native Hen (*Tribonyx mortieri*) has been called cooperative polyandry, to distinguish it from the polyandry of a number of shorebirds and other ground-nesting, precocial species, among which the female provides sets of eggs for several males, each of whom incubates and attends the chicks with no help from her or her other consorts. Another raptor that practices cooperative polyandry is the Harris' Hawk (*Parabuteo unicinctus*), a blackish bird with chestnut shoulders, thighs, and wing linings and white upper and under tail coverts. This inhabitant of arid regions is found from the deserts and semideserts of southwestern United States through much of Middle and South America to central Chile and Argentina. Amid the cacti, mesquite, and palo verde of the Sonoran Desert of southern Arizona, William Mader discovered a strongly unbalanced sex ratio, with 2.8 males to 1 female. Nearly half of fifty nests had three attendants.

At the three nests where Mader learned the birds' sexes, the trio consisted of two males and a female. Both males copulated with the female and took turns with her incubating one to four, usually three, unmarked white eggs in the raptor's typical nest of coarse sticks in a giant cactus or a tree. When, after about thirty-five days of incubation, the nestlings hatched, both male parents brooded them and brought food for them and their mother. Largest and the dominant member of the trio, she caught no prey but was nourished by the males while she performed the major part of diurnal incubation and all nocturnal incubation, brooded and shaded the chicks, and delivered to them food that her partners brought. The three adults lived in concord, exchanging food or sharing the same mammalian, avian, or reptilian victim. In contrast to the chicks of many other raptors, who often aggressively deprive their siblings of food or even kill and devour them, nestling Harris' Hawks dwelt in an amity that presaged their peaceful cooperation as nest attendants in later life.

Trios had substantially higher nesting success than pairs, raising, on the average, 2.5 fledglings per successful nest, while pairs raised only 2.2. However, pairs produced 1.1 fledglings per adult, trios only 0.83. As in similar cases, the advantage to the female of polyandry was clear: she raised more young with less effort. In view of the paucity of females, males doubtless did better by joining in a polyandrous association as soon as possible than by waiting long for an unattached female or possibly passing their whole lives as bachelors. Especially in a harsh, unpredictable climate like that of the Sonoran Desert, one more forager to provision the nest, with no increase in the number of nestlings, must often save the young from starvation.

As in other orders of birds, polygamy, or a nest helper, is occasionally found in species that normally breed in unaided monogamous pairs. For nearly a month, two male American Kestrels, or Sparrow Hawks (*Falco sparverius*), brought food to a female with nestlings in New York State. Whether this was a polyandrous association or whether the second male was a nonbreeding helper was not ascertained. In Africa, an immature Martial Eagle (*Polemaetus bellicosus*) incubated and fed nestlings after the original adult male disappeared. In California, a male and two female Red-tailed Hawks (*Buteo jamaicensis*) attended a nest with four eggs, probably laid by only one of them. The two females took substantial shares in brooding the nestlings, while the male brought food to both of them. No antagonism was observed among the three nest attendants. This was the only one of fifty-three active Red-tail nests in which four young were raised.

On the coast of Massachusetts, Ospreys (*Pandion haliaetus*) nested on high platforms erected for them on open salt marshes. After they could fly, young from one nest sometimes joined the brood on another platform in view of the first, where they shared food brought for the resident nestlings. Usually it was the smallest and apparently youngest member of a brood, disadvantaged in the competition for food, who abandoned its natal nest to visit a neighboring nest with a younger brood where it was no longer subordinate. The parents were unexpectedly indulgent of such intrusions.

The unnatural proximity of birds of diverse species and ages in an aviary sometimes encourages activities that are unlikely to arise among free birds. In Kenya, a captive fledgling Black-shouldered Kite (*Elanus caeruleus*) named Elanus brooded and fed chicks of her own species, rearing them from the age of one day to independence. She also adopted a day-old buzzard chick and continued to feed it for nearly two months, until it was thrice her own size. She guarded a duckling and "brooded" a red notebook and other inanimate objects of the same color. In Wisconsin, a juvenile Red-shouldered Hawk (*Buteo lineatus*), sixty-five to seventy days old and still dependent for food on the Red-tailed Hawk that reared her in captivity, tried repeatedly, without success, to feed young Northern Harriers, or Marsh Hawks (*Circus cyaneus*), through the wires of their cage. She also tried to build a nest with green leafy twigs. With abundant, readily accessible food and reduced opportunities for flight and exploration, captive birds often engage in parental activities long before they are mature.

References: Acker 1977; Faaborg et al. 1980; Mader 1975, 1976, 1979; Poole 1982; Tarboton 1976; Van Someren 1956; Wegner 1976; Wiley 1975.

5.

RAILS, COOTS, AND GALLINULES

Rallidae

Distributed over most of the Earth, except the polar regions, the 129 species of the rail family range in length from 5 to 26 inches (13 to 66 centimeters). Most have narrow bodies that easily slip through dense vegetation, short wings, and long legs and toes. Shy and secretive, protectively clad in browns, grays, and black, frequently barred, streaked, or spotted, rarely brightly colored, rails live obscurely in marshes, swamps, and dense stands of grasses or reeds and are difficult to know intimately, which is a pity, for their habits are exceptionally interesting. They sing duets, build special platforms or nests for sleeping or for brooding their downy chicks above water or wet ground, and in their bills carry to greater safety eggs or young that appear to be in danger. They eat a large variety of vegetable and animal foods. Rails build open or roofed nests and lay from two to sixteen white or buffy eggs, spotted and blotched with shades of brown and pale lilac, that are incubated by both sexes.

One of the less elusive members of the family is the Common Moorhen (*Gallinula chloropus*), in America formerly known as the Common, or Florida, Gallinule. Over much of the Americas as well as Eurasia and Africa, it inhabits freshwater ponds, marshes, and the grassy or weedy margins of larger bodies of water. With brownish back and wings and blackish head, neck, and underparts, its brightest color is the red of its frontal shield and of all except the yellow tip of its bill. The drabness of its plumage is relieved by a white horizontal streak along each side and white under tail coverts. A plump, henlike bird about 12 inches (30 centimeters) long, the Common Moorhen swims with bobbing head, flies clumsily, and climbs through waterside vegetation.

Where protected in England, moorhens become tame and confiding. Viscount Grey of Fallodon gave a charming account of a family of free birds that nested on a pond in his garden. In May the parents hatched a brood, to which they gave bread that he threw them. In mid July, when the old birds had brought forth a second brood, three survivors of the earlier brood, fully feathered and well able to feed themselves, remained with them. Now, when Grey tossed bread to the parent moorhens, they would pick it up and put it into the beak of one of the May young, who in turn would pass it to one of the tiny downy chicks of the later brood. This was repeated again and again. Sometimes a May bird, after receiving a piece of bread from a parent, passed it to another May bird, who in turn gave it to a downy chick. Once a parent fed a chick directly, but this was apparently against the rules, for a young helper promptly ran up, snatched the bread from the bill of the infant bird, then replaced it there.

The juvenile moorhen's habit of feeding chicks is widespread. In early July, on a weedy pond in the Costa Rican highlands, I found a family consisting of two adults, two full-grown young birds in grayish plumage with dark bills and foreheads, and four downy chicks who kept up a constant peeping. The juveniles seemed to give as much attention to them as the parents did, and once one of the former appeared to pass food to a chick. Similar behavior has been reported from Louisiana and Pennsylvania and from South Africa, where chicks begin to feed themselves when one week old and are capable of supporting themselves at three weeks, although the parents normally continue to provide some food for them for another three weeks. Juvenile helpers from four to ten weeks old attend and feed their younger siblings, but "the frequency and magnitude of this care probably result in no great assistance to the parents."

Attired in purplish blue and bronzy green, with white under tail coverts, a pale blue frontal shield, red bill tipped with yellow, and long yellow legs, the Purple Gallinule (*Porphyrula martinica*) is an exceptionally colorful member of the rail family. Less widespread than the Common Moorhen, it is found only in the western hemisphere, from Tennessee, Ohio, and Maryland through southeastern United States, over much of Central and South America to northern Chile and northern Argentina, as well as in the West Indies.

On a shallow, 7-acre (3-hectare) pond, largely overgrown with vegetation, near Turrialba, Costa Rica, Charles Krekorian found six families of Purple Gallinules, four of which contained both juveniles and chicks. Starting when at most sixty-nine days old, the juveniles fed the downy young, each of those most carefully watched giving them food about as frequently as each adult did. The chicks continued to be fed by the juveniles until the former were about fifty days old and had long been feeding themselves. The principal food of the Purple Gallinules on this pond was the fruit of the broad-leaved water lily (*Nymphaea*), but their fare also included stingless bees, frogs, grass seeds, spiders, worms, and fish. Although these gallinules appeared to be predominantly monogamous and the juvenile helpers appeared to be older siblings of the chicks, in a family in which the chicks were fed by four or more adults as well as by juveniles, relationships were less clear and suggested the possibility of diverse breeding systems. In addition to feeding and protecting the downy young, juveniles helped adults defend the family's territory.

A different system, combining communal nesting with cooperative breeding, is followed by the Pukeko, or Swamp Hen (*Porphyrio porphyrio*), in marshes and wet pastures of New Zealand. In a three-year study, John Craig found this gallinule living in monogamous pairs or promiscuous groups of three to six birds, on territories defended by all group members, including juveniles. Equal numbers of males and females composed these groups, which tended to be stable and to remain on their breeding territories throughout the year. In each group or pair with young, a lineal dominance hierarchy was maintained, with males dominant over females and older birds dominant over younger ones. Subordinate members preened dominant members slightly more often than they were preened by them, but dominant birds fed their subordinates more frequently. After the breeding season, part of the population joined in flocks of, rarely, as many as 120 individuals. At the next nesting season, some of these flocking Pukekos formed new breeding groups, with a preponderance of males and an unstable membership.

Craig saw only males building nests, open bowls of interlaced leaves supported by the stems of emergent cattails or rushes or lying in tussocks of sedges. Each nest was screened by overhead vegetation and approached by a ramp. About a month before laying began, a number of nests were started and left incomplete, but finally two or three were carefully finished. In one of these, the dominant female laid from four to six eggs, and in about two-thirds of these nests the subordinate female deposited a usually smaller number. Rarely, three hens contributed to a composite clutch. Two eggs were frequently deposited on the same day, and those of each female could often be distinguished by differences in color or size.

During an incubation period of about twenty-five days, only adult Pukekos warmed the eggs. The dominant female did most of the daytime incubation, the subordinate female slightly less. The two males participated equally but took fewer turns on the eggs. A male, nearly always the dominant one, sat at dusk and was relieved immediately before dawn by the dominant female, who was in turn replaced by the subordinate female. The dominant male was usually the first to return to the nest after a disturbance, especially by people. Nonbreeding yearlings rarely visited a nest and were not seen to incubate. In solitary pairs, the male and female regularly alternated on the eggs for diurnal sessions of approximately three hours.

While the eggs hatched, one parent continued to cover the nest, and group members either fed the chicks directly or gave food to the sitting bird, who passed it to the downy ones. Within three days after they hatched, the chicks were led from the nest and fed nearby, often on platforms built on the water with many leaves. All group members guided, protected, and fed the young Pukekos. When nonbreeding yearlings were not present, the dominant male cared for the chicks substantially more than any other member of his group, the subordinate female was next, and the subordinate male did little. When nonbreeding yearlings were present, they contributed greatly to chick care, a yearling male often doing more than any other group member. Although the downy young began to feed themselves when only two days old, they continued to receive much food from the adults until they were two months old. About the time they hatched, special nests were built for brooding them at night, or they might sleep beneath a parent on an old breeding nest.

Pukekos often rear two broods in a season, and juveniles of the earlier brood help feed their younger brothers and sisters. When a pair nested alone, young of the first brood, forced to become independent at an earlier age than young birds on group territories, contributed about half the care of the second brood. Unlike larger groups, solitary pairs lack yearling helpers because their offspring stay with them only until nesting begins in the following season.

Despite their lack of helpers older than juveniles, pairs bred more successfully than larger stable groups, producing two or three times as many surviving young per pair per season and from three to six times as many per adult per season. Unstable groups, newly formed of birds from the winter flocks, failed in all their attempts to raise young. The greater productivity of the solitary pairs, exceptional among cooperative breeders, raises the question of the value of the auxiliaries. Do the undoubted services of helpers who feed and protect chicks count for nothing? Are Pukeko helpers actually counterproductive? From his painstaking study, Craig concluded that to assess the value of cooperative breeding in the Pukeko, one must consider more than the success of the reproductive effort. Many variables, including the size and quality of the territory, the age and experience of the breeding adults, and the duration and stability of their association, affect the outcome of nesting in ways difficult to disentangle.

Although monogamy appears to predominate in the rail family, the flightless gallinule known as the Tasmanian Native Hen (*Tribonyx mortieri*) is definitely polyandrous, a situation related to its unbalanced sex ratio of three males to two females. This "oversized" greenish brown moorhen, confined to the southern island, has a white patch on each flank, a greenish yellow bill, red eyes, and stout legs with which it can run at the rate of 30 miles (48 kilometers) per hour. Its preferred food is tender grass shoots, which it finds chiefly in low, moist areas grazed by mammals—in former times by native Tasmanian marsupials, now chiefly by introduced sheep. Unfortunately, the moorhen's liking for the young sprouts of such lawnlike swards and of cereal crops has brought it into conflict with farmers, who have advocated its extermination, although the extent of the damage attributed to it appears to have been exaggerated. In the drier season it supplements its diet of green herbage with seeds, chiefly of grasses. Insects are the main food of recently hatched chicks.

In a three-year study of banded native hens, Michael Ridpath found them living on vigorously defended territories in groups of up to seventeen, including young who stay with their parents for nearly twelve months. Considering only adults, who tend to remain together year after year, fifty-one groups included twenty-six pairs, nineteen trios, five quartets, and one quintet. Trios usually consist of two brothers mated with a female from some other family, rarely of brothers with a sister. Although strenuous fights arise between territory holders and invaders, members of a breeding group dwell in great amity, with little evidence of the dominance hierarchies frequent among birds and other animals. In late winter or early spring, the single female present in most groups lays five or six eggs in open nests built by both sexes on the ground, beneath dense cover near water. Parents of both sexes incubate the eggs for about twenty-two days, then feed the chicks, often helped in all these activities by other members of the larger groups. Chicks start to feed themselves when a week or two old but continue to receive food from older group members until they are about two months old.

Far above the tree line in the cold, bleak puna zone of the Andes, from central Peru to northern Chile and northwestern Argentina, the Giant Coot (*Fulica gigantea*) lives on shallow lakes 11,810 to 16,400 feet (3,600 to 5,000 meters) above sea level. A slaty black bird 26 inches (66 centimeters) long, this coot has a dark crimson, white-tipped bill, a yellow frontal shield that is white in the center, and red legs and feet. In shallow water well out from the shore, it builds floating nests of aquatic plants; with yearly increments, the nests become huge structures up to 10 feet (3 meters) long at the water surface that finally rest on the bottom. The high rims of these "island nests" not only serve to protect the chicks from the chilling gales that sweep across the puna but, with the water plants that the parents continue to deposit upon them, become a source of food for the chicks, who before they are two weeks old try to feed themselves from this accumulation. Parents, helped by juveniles from earlier broods, continue to feed the chicks until they are about two months old. Likewise, among European Coots (*F. atra*), juveniles feed chicks. Adults adopt the lost young of other parents.

Another form of precocious parental behavior widespread in the rail family is nest building. Immature individuals of the Common Moorhen and of the Black Crake (*Limnocorax flavirostris*) of Africa have been observed helping build or repair nests. A captive three-week-old Sora (*Porzana carolina*) picked up and carried pieces of dead grass; and a month-old American Coot (*Fulica americana*) was watched throwing grass over her shoulder, accumulating a little pile by a method widespread among birds that build simple nests on the ground.

References: Alley and Boyd 1950; J. L. Craig 1979, 1980a, 1980b; Fjeldså 1981; Grey of Fallodon 1927; Krekorian 1978; Nice 1962; Ridpath 1964; Rowley 1975; Siegfried and Frost 1975.

6.
PIGEONS AND DOVES
Columbidae

The 284 species of pigeons and doves are found in all tropical and temperate regions of the Earth. In size they range from the smallest doves, about 6 inches (15 centimeters) long, to the crowned pigeons of New Guinea, about 30 inches (76 centimeters) in length. In coloration they are extremely diverse, some being rather plainly attired in grays or browns, while others are gorgeous with red, orange, yellow, green, blue, or purple. A few have high, upstanding crests. The most elegant pigeons live in southeastern Asia and Indonesia and on islands of the southwestern Pacific. The sexes may be alike or quite different. Some forage in high treetops, others on the ground, taking fruits, seeds, and insects. Their song is often a simple cooing, but a few achieve a pleasing complexity.

To build their artless open nests, in trees or shrubs, on the ground, or on cliffs, the female pigeon usually sits on the site and arranges the twigs, straws, or other materials that her mate delivers to her, one by one. The clutch usually consists of two white or buffy eggs, frequently only one, very rarely three or four, perhaps the product of two females. Incubation follows a stereotyped pattern, the male sitting through many of the daylight hours, the female all the rest of the time. At first nourished with "pigeon milk," regurgitated from the crops of both parents, the nestlings receive increasing amounts of solid food, also regurgitated, as they grow older. The tight incubation schedule and the special diet of milk secreted only by parents who have incubated seem hardly compatible with cooperative breeding such as we find in birds who can share with others whatever food they select. Nevertheless, pigeons are not devoid of helpful impulses, as the following examples attest.

The Wood Pigeon (*Columba palumbus*) is abundant and familiar in the British Isles, where it builds its nests on ledges of buildings in towns as well as in trees and shrubs in rural areas. Its two eggs hatch in fifteen and a half to seventeen days, and its young remain in the nest for about twenty-two days. After feeding its two nestlings, a parent flew to and fed a strange juvenile who begged for food on a perch about 10 yards (9 meters) away. At three different nests with small nestlings, an intruding juvenile was fed. On four consecutive days, a juvenile flew from one of these nests when visits were made to weigh the nestlings, but later it returned. Such indulgence is not invariable, for at this stage juveniles are not always tolerated by adults, who may attack and drive them from their territories.

A fledgling domestic pigeon or Rock Dove (*C. livia*), about twenty-five days old, regurgitated food to a younger companion. Another young pigeon, about a month old, took sticks to the nest where his mother incubated and presented them to her as an adult male does to his mate. After some initial difficulties, he daily incubated his mother's eggs for about two hours at a stretch.

In southern Arizona, a Mourning Dove (*Zenaida macrura*) brooded and fed nestling White-winged Doves (*Z. asiatica*), a few days old, who had been neglected by their own parents through much of the day. When at last the female White-wing returned, in the late afternoon, she fought and drove away the helpful Mourning Dove. The latter, nevertheless, continued to minister to the young White-wings until they fledged. Apparently, her own eggs had failed to hatch. In aviaries, Mourning Doves and several other kinds of doves are quick to adopt and aid in the care of young pigeons of any species. In large colonies of White-winged Doves in southern Texas, helpers sometimes feed nestlings. See, also, the accounts of the Yellow-billed Cuckoo (*Coccyzus americanus*) (chapter 8) and the American Robin (*Turdus migratorius*) (chapter 36).

REFERENCES: Goodwin 1947; Murton and Isaacson 1962; Neff 1945.

7.

PARROTS

Psittacidae

The 332 living species of parrots are conspicuous in almost all tropical regions except treeless deserts, with many extending into the subtropics and a few in colder climates. In size they range from pygmy parrots (*Micropsitta*) scarcely over 3 inches (8 centimeters) long to Hyacinth Macaws (*Anodorhynchus hyacinthinus*) 40 inches (1 meter) in length. Their plumage is prevailingly green, but all colors, as well as black and white, are generously displayed by this brilliant family. Their food is largely fruits, seeds, and nuts, which they crush with their thick, powerful bills, and many of the smaller parrots drink nectar. Their usually raucous voices may sound melodious at a distance. Most nest in unlined holes that they find ready-made in trees, in chambers that they excavate in termitaries, in burrows in cliffs, or, rarely, as with certain lovebirds (*Agapornis*), in domed nests that they build in sheltered places. One to about a dozen white eggs are nearly always incubated by the female, who is fed by her mate; in a few species he shares incubation. Both parents regurgitate food to the young.

On present information, only two quite different species of this great family are cooperative breeders, and for one the evidence is far from adequate. This is the Eclectus Parrot (*Eclectus roratus*), a stout, thick-billed bird 14 inches (36 centimeters) long. The males are largely green; the very different females are red, ornamented with purple and yellow. They live in lowland forests or on savannas with clumps of tall trees on many of the islands of the southwestern Pacific, from the Moluccas to the Solomons, and on the Cape York Peninsula of Australia. The parrots roost gregariously in trees and, in the morning, fly screeching and squawking high above the forest canopy to seek fruits, nuts, seeds, flowers, and nectar in the treetops. In cavities high in the trunks of tall trees near the edge of the forest or in a clearing, the female lays two eggs on a bed of wood chips. During incubation, her mate calls her from the nest to feed and preen her. From all parts of the species' range come reports of up to seven or eight Eclectus Parrots of both sexes attending a nest. Four males and two females were present at a nest that Joseph Forshaw found in extreme northern Australia. Since we are not told what they did there, we must suspend judgment as to their status as helpers pending further information. Parrots are notoriously difficult to study in the wild because their nests are so often high and inaccessible and the birds themselves are shy and suspicious. Much that we know of their breeding comes from aviaries. A captive female Eclectus incubated two eggs for twenty-six days, leaving the nest only twice a day to be fed by her mate. The young bird fledged when slightly over twelve weeks old.

A bird whose nesting habits are unique in its family, the Monk Parakeet (*Myiopsitta monachus*) inhabits open woodlands, gallery forests, savannas with scattered trees, farmlands, and orchards from Bolivia, Paraguay, and southern Brazil to central Argentina. It has been introduced to Puerto Rico, where it has become established, and to the United States, where it spread widely but sparsely through the eastern half but possibly will not remain because efforts have been made to exterminate it as a threat to crops. A long-tailed bird nearly 12 inches (30 centimeters) in length, it is largely green, with a gray forehead, cheeks, throat, and breast. On the grassy pampas of Argentina, where trees for nests were originally scarce, extensive planting since the arrival of Europeans has greatly increased suitable sites and favored the multiplication of Monk Parakeets, until their liking for cereal crops, as well as a variety of fruits, nuts, and flowers, made them a menace to agriculture and provoked the Argentineans to resort to harsh measures to destroy them.

Probably because of the scarcity of nest cavities in regions with few massive tree trunks, Monk Parakeets evolved the habit of building in the open, and because trees of any kind were rare, they crowded their nests together on the few suitable ones. An apartment nest is begun when a single pair builds, of tightly interlaced, often thorny twigs, a bulky structure with an entrance tube near the bottom, leading upward to a porch or antechamber and a nest chamber. Other pairs add similar structures, until the nest mass becomes over 6 feet (2 meters) high and may weigh a quarter of a ton (225 kilograms). The largest masses may contain as many as twenty compartments, but those with more than a dozen are rare. The rooms do not intercommunicate; each has its own entrance. Surprisingly, these huge nests are not supported on massive limbs but hang from the ends of branches or drooping palm fronds, about which they are firmly woven. Some old trees have seven or eight such masses suspended from their boughs. Throughout the year, the parakeets keep their nests in good repair and sleep in them, but they build new chambers only as the breeding season approaches. Then each female lays from five to eight eggs on the unlined floor of the chamber.

The degree of cooperation by Monk Parakeets is not clear, as details of their breeding behavior are lacking. The parakeets do not, like Sociable Weavers (*Philetairus socius*), build a common foundation or roof to which all the chambers are attached—each pair appears to build and maintain its own compartment—but without some cooperation to keep the whole conglomerate in a sound state it might fall apart. An obvious mode of cooperation among close neighbors is the noisy harrying of dangerous raptors. However, the parakeets permit birds as large and heavy as Jabirus (*Jabiru mycteria*), Chimangos (*Milvago chimango*), and whistling-ducks (*Dendrocygna*) to raise their broods on top of their nest masses.

REFERENCES: Conway 1965; Forshaw 1977; Hudson 1920.

8.

CUCKOOS AND ANIS

Cuculidae

Distributed over most of the Earth except the coldest regions, the 127 species of the cuckoo family range in length from 6 to 26 inches (15 to 66 centimeters). Some are shining, metallic green, others somber black, and between these extremes of coloration are grays, browns, yellows, and white, often with bars or streaks. Most cuckoos are arboreal inhabitants of forests and thickets, but many are strong-legged ground foragers. Although the family is largely insectivorous, with a taste for hairy caterpillars, the terrestrial species capture many lizards, snakes, and even small mammals and birds. Cuckoos that nest at high latitudes undertake long migrations, sometimes to small tropical islands difficult to find in vast expanses of ocean. Cuckoos' nests vary from flimsy constructions of twigs to the substantial domed structures with a side entrance, lined with green leaves, of the coucals that range from Africa to Australia. Their eggs are often white, greenish blue, or blue; but those of some of the parasitic species vary widely in color to match the eggs of their different hosts. Many cuckoos breed in monogamous pairs, of which both sexes incubate and care for the young; but the family includes forty-two species of nest parasites in the Old World and three species in tropical America. Four species in the subfamily Crotophaginae nest communally.

Most thoroughly studied of the communal nesters are two of the three species of long-tailed black anis that, alone or together, are found over most of the tropical and subtropical regions of the western hemisphere. The Smooth-billed Ani (*Crotophaga ani*), distinguished by its compressed, high-arched upper mandible, inhabits South America from northern Argentina northward, Panama, the southern Pacific quarter of Costa Rica, the West Indies, and southern Florida. The slightly smaller Groove-billed Ani (*C. sulcirostris*), whose less strongly arched upper mandible bears prominent longitudinal ridges and furrows, is less widely distributed in South America than the Smooth-billed but occurs alone over the greater part of Central America and Mexico and in southern Texas. As similar in their social habits as they are in appearance, the two anis may well be considered together.

Anis are birds of wide ecological tolerance, residing from warm lowlands, where they are most abundant, up to 7,000 or 8,000 feet (2,135 to 2,440 meters) in cool highlands and in dry as well as very rainy regions. Although they avoid heavy forest, they seem equally at home in clearings in rain forests, dry savannas, upland pastures, plantations, gardens, and marshes and amid cacti and thorny scrub. In loose flocks of about ten to twenty-five individuals that vigorously defend their territories against neighboring groups, they hunt insects and spiders among the foliage of trees, shrubs, and vine tangles but chiefly in grassy or weedy fields, over which they run or hop with feet together, often leaping into the air to seize some flying insect. Their most productive foraging occurs close to the heads of grazing quadrupeds—cattle, horses, or mules—who stir up grasshoppers and other insects that are readily caught by the anis. When army ants extend their raids beyond the forest, these birds accompany them to capture small fugitive creatures. Although they are reputed to pluck ticks from cattle (a reputation responsible for the name *garrapateros* or tick eaters given to them in several Spanish-speaking countries), this has not been confirmed by the most careful observers. Especially in dry weather, anis diversify their diet with berries.

Anis are highly social, affectionate birds who live in great amity with group members. If one becomes separated from its comrades as they straggle through pastures or open groves, it calls and calls until reunited with them. To rest by day, these black birds line up as close together as they can press on a horizontal branch; and they sleep, preferably in a thorny tree or amid dense foliage, in the same compact formation. They carefully preen the plumage of companions who may or may not be their mates. As the nesting season approaches, monogamous pairs keep close company without separating from the flock.

In regions with even a light dry season, anis nest later than many of their avian neighbors, as they wait until returning rains have promoted a lush growth of the herbaceous vegetation among which grasshoppers and other invertebrates flourish. For their nest site they prefer a thorny tree with dense foliage, often an orange or some other kind of citrus growing in an open space where they can forage, a vine-draped tree, or a dense tangle of creepers at the margin of a thicket. In arid regions they may choose an organ cactus or opuntia bristling with needlelike spines. In height their nest sites range from low amid marsh vegetation to the top of a tall tree, but most are between 5 and 20 feet (1.5 to 6 meters) above the ground. Cuckoos in general are not among the most skillful of builders, and anis are no exception to the rule. Their nests—untidy but sturdy open bowls of interlaced coarse twigs, inflorescence branches,

Groove-billed Anis

weed stems, and the like—are lined with leaves brought while green. To build, they follow, more or less closely, the method widespread among pigeons: the female sits on the incipient structure to arrange materials that her mate brings to her. His twigs may be broken from trees or shrubs or gathered from the ground; leaves are pulled from living branches. This method is followed not only by solitary pairs but in large measure by communal nesters, for anis prefer to work with their mates.

Breeding groups tend to be smaller than off-season flocks, which may include immature birds. Nests may be attended by solitary pairs or by as many as fifteen individuals. The relative abundance of pairs and communal groups varies with the habitat. On the arid Santa Elena Peninsula of western Ecuador, only five of twenty-four nests of Smooth-billed Anis appeared to belong to more than one pair. In northwestern Costa Rica, single pairs of Groove-billed Anis predominated in pastures, whereas six was the most frequent number of attendants at nests in the lusher vegetation of nearby marshes. In any case, groups of more than six or eight breeding adults are rare because nests with too many eggs are inefficient.

Unlike songbirds, which usually lay early in the morning, anis deposit their eggs chiefly around noon or in the early afternoon, at two-day intervals, although occasionally they do so at almost any hour of the day. Freshly laid eggs are an immaculate white, covered with a thin, chalky layer that is easily scratched off by one's fingernails or the birds' bills or toenails, revealing the blue or blue-green of the underlying shell. Where not scraped away, the white layer soon becomes stained by contact with the bed of green leaves. A single female may lay from three to eight eggs, but four or five is usual. Thus, nests of solitary pairs tend to have four eggs, while those belonging to two pairs contain eight or nine, and those with three laying females most often have ten to twelve. The largest set reported for the Groove-billed Ani contained twenty eggs, attributed to five breeding females. A nest of the Smooth-billed Ani with twenty-nine eggs belonged to a group of fifteen birds. An old report of a Smooth-bill nest with 151 eggs, arranged in layers separated by leaves, is hard to reconcile with more recent studies. In nests with more than a dozen eggs, some tend to remain buried beneath the green leaves that are added daily during incubation and never removed, and incubation is inefficient. Eight was the greatest number of eggs hatched by Smooth-billed Anis in Cuba, where many sets were much larger; and in Costa Rica thirteen was the maximum number hatched in any nest of Groove-billed Anis.

Sandra Vehrencamp discovered that at communal nests in Costa Rica the oldest, dominant, or alpha female, mate of the alpha male, was the last to lay—and, before depositing her first egg, she tossed all eggs already present from the nest. Other females continued to lay after the loss of their early eggs, replacing some of them, but each rarely contributed as many incubated eggs as the dominant ani had. It is surprising to find such unfriendly, wasteful behavior among birds who otherwise live in great concord; but, in addition to giving the alpha female an advantage over her associates in the number of her progeny, egg tossing may benefit the whole group because it shortens the interval during which incubated eggs are laid, with the result that they hatch in a briefer period and the nestlings, of a more uniform age, will all have better chances of surviving. Such egg ejection has not been reported by other observers of anis, but the intact or broken eggs not infrequently found beneath nests that have apparently not suffered predation may well have been thrown out by some of the attendant birds.

In the days following the completion of their set, anis gradually increase the constancy of incubation until, during the latter part of the incubation period, they keep their eggs continuously or almost continuously covered. For nests with two or more attendant pairs, this entails no great effort by individuals; all the breeding birds take turns on the eggs, although the time devoted to incubation by the individuals of the group may be very unequal. At a nest of Smooth-bills in Colombia belonging to three pairs and three unmated, probably immature birds, the latter also incubated, but two of them spent very little time on the eggs because, when they came to sit, they were rebuffed by breeding females already present. The third unmated ani was more tolerated by older group members. Not every member of the group incubated every day. Just as the contributions to incubation by the several collaborators are very unequal, so are the lengths of their individual sessions, which may vary from a minute or two to well over an hour. Any disturbance of the nest is likely to cause very frequent changeovers. Just after I had entered a blind before a nest of Groove-billed Anis, all four attendants took turns on the eggs in less than ten minutes. As in certain other cuckoos, woodpeckers, and many other birds, nocturnal incubation is the male's duty. Apparently, the single ani who sits through the night is regularly the alpha male, who is also the most assiduous incubator by day.

The nesting activities of anis do not follow the strict sequence that one expects of passerine birds. They bring leaves for the lining before their foundation of sticks is completed, and they may even lay eggs in an unfinished nest. During incubation, they continue to bring sticks and green leaves to the nest. If they do not wish to incubate, they deliver their contribution to the partner already sitting there and fly away, while the recipient arranges the stick in the framework or tucks the leaf beneath the eggs. If the ani arriving with stick or leaf wishes to sit, it waits beside the nest until the bird already present arises and goes, then enters the nest and arranges its own contribution. At a nest of Smooth-bills in Colombia, females brought many more leaves than twigs, while males brought substantially more twigs than leaves. If the bird covering the eggs is reluctant to depart, the newcomer may sit beside it or even push beneath it, whereupon the other goes. Two anis apparently never incubate side by side for more than a minute. The bird so displaced, perhaps after having sat very briefly, never reveals displeasure or resentment.

After an incubation period of approximately thirteen days, counting from the laying of the last egg to its hatching, the nestling anis, blind, black-skinned, and devoid of down, escape from their hard shells. Their gaping mouths reveal an intricate pattern of white marks on a red ground. Because of irregularities in the time of laying and the start of incubation, hatching of a large set of eggs in a communal nest may be spread over several days, with the result that the youngest nestlings, much smaller than their rapidly growing nest mates, are at a great disadvantage and may starve or be trampled—but such losses are not infrequent in large broods of birds that do not nest communally.

As the nest attendants incubated by turns, so now they brood the nestlings by turns through the day, while at night the alpha male covers them. The young are fed not only by the parents and adult helpers but, if they belong to a second brood, by juveniles from the earlier brood. A Smooth-billed Ani started to feed nestlings when forty-eight days old, a Groove-bill at seventy-two days. The nestlings are nourished chiefly with insects, supplemented by an occasional small lizard. After two attendants have tried strenuously but unsuccessfully to pull apart a lizard between them, it may be delivered entire to a nestling, who swallows it headfirst and rests with the long tail projecting from its mouth while the reptile's foreparts are slowly digested. Adults swallow the droppings of very young nestlings, but soon the latter try, not too successfully, to eject their excreta beyond the nest's rim, thereby fouling it and giving the nest a characteristic odor. Parents, especially males, defend their young with spirit. Voicing a menacing *grrr* and snapping their strong bills with a loud clack, Groove-bills have repeatedly buffeted the back of my head when I visited their nests. Juveniles a few months old join these defensive demonstrations.

Nestling anis are extremely precocious. When only six days old, with feathers just emerging from the ends of their long, horny sheaths, they jump from a nest that is disturbed, to climb and hop away through the surrounding branches, hooking their bills over twigs to catch themselves when a leap falls short of its goal. If one drops to the ground, it creeps into the herbage and hides. Later, when the excitement has died away, these fugitives return to the nest, if they are able, to be brooded through the night until they are about eight days old. For several days more, they are half-scansorial, half-terrestrial, climbing and hopping with agility through close-set branches or creeping through the ground cover between shrubs and trees. At eleven days they are well feathered and can make short flights from branch to branch of the same tree. Everywhere black as their parents, they have short tails, bills devoid of grooves, and dark beady eyes, shaded by eyelashes, set amid bare black skin.

Aside from satisfying their strong social impulses, what advantages do anis derive from nesting communally? To answer this question, Sandra Vehrencamp studied the nesting success of groups of different sizes occupying diverse habitats in semiarid northwestern Costa Rica. She found that the number of attendants per nestling made little difference in the rate at which the young were fed; whether there were more grown birds than nestlings or more nestlings than attendants, the latter adjusted their feeding visits to the young birds' needs. In pastures, where diurnal predators were active, larger groups defended their nests more effectively than did smaller groups; but in the marsh, where much of the predation occurred by night when anis are helpless, group size did not affect the rate of predation. When allowance was made for the effects of different habitats, the annual reproductive success was the same for small as for large groups.

Although communal nesting did not directly increase the production of young, it did influence adult survival, particularly that of the subordinate males and the females, who did not incur the high risk of nocturnal incubation. Alpha males, who alone were in charge of communal nests by night, suffered the highest mortality. Not only does communal breeding, by decreasing the number of nests in a population, diminish the number of males who incubate at night, but by dividing diurnal incubation among a number of collaborators, it increases the time that all members of the group enjoy for foraging and other aspects of self-maintenance. It appears that among anis the chief advantage of communal nesting accrues directly to the participating adults rather than to their rate of reproduction. Indirectly, it does increase reproduction, for by living longer anis can raise more broods; and subordinate males, who father fewer nestlings because their mates' eggs are ejected by the mate of the alpha male, may by surviving the alpha male compensate for their losses.

The third species of *Crotophaga*, the Greater Ani (*C. major*), is about 17 inches (43 centimeters) long, much larger than the other two. Glossy blue-black, with a long purplish tail and bright yellow eyes, it is also much more handsome. The high, compressed ridge at the base of the upper mandible gives its bill a peculiar, misshapen aspect. In pairs or groups of up to ten or twelve, it inhabits swamps, riverside and lakeside thickets, mangroves, and wet areas in general, at low altitudes in central and eastern Panama and over much of South America to northern Argentina. A noisy bird, the Greater Ani utters deep croaks, loud musical monosyllables, and, often in chorus, a series of bubbling notes that have earned it the vernacular name *hervidor*—boiler. Its diet is much like that of the other anis. Like them, it builds, usually near water, a cup of twigs lined with leaves, which may contain ten or more eggs with a white, chalky deposit covering the hard blue shell. In Trinidad, four adults attended three fledglings, but no thorough study of the Greater Ani's communal nesting has been published.

A relative of the anis is the Guira Cuckoo (*Guira guira*), which ranges from islands in the mouth of the Amazon through eastern Brazil to the Río Negro in south central Argentina. The Guira is about 16 inches (41 centimeters) long, with a rufous crest. Its upper back and wing coverts are dark brown, streaked with white. The lower back and underparts are buffy white, narrowly streaked with black on the chest. The central feathers of its long tail are dark brown, the others dark glossy green, with yellowish bases and broad white tips. The eyes are red, the bill orange-red, and the legs blue. No less social than the anis, it lives in parks and open country with scattered trees in flocks of often a dozen to twenty individuals, which roost not only pressed side by side but on the backs of others in two or even three tiers—at least beyond the tropics in Argentina, where, according to W. H. Hudson, these thin-plumaged birds of tropical origin suffer greatly from winter's cold and sometimes fall frozen from their roosts. They spend much time basking in the sunshine to warm their chilled bodies, and they preen each other.

The Guira Cuckoo's nest, a rough open cup of sticks lined with green leaves or grasses, is built in a thorny tree or amid dense foliage and vines. Its eggs differ from those of anis in that the beautiful turquoise-blue shell is sprinkled with delicate white flakes which are readily brushed off, instead of being completely covered with white. At one time, a single group may have several nests, some belonging to solitary pairs who lay five to seven eggs, others with up to twenty-one eggs attended by larger groups. In addition to eggs in nests, many are scattered over the ground. Guira Cuckoos are apparently less socially advanced than the anis, but we need more thorough studies of their habits, especially in the tropical parts of their range.

The two species of cuckoos widespread in temperate North America, the Black-billed (*Coccyzus erythropthalmus*) and the Yellow-billed (*C. americanus*), often deposit eggs in each other's nests, likewise in nests of other small birds, possibly because their own poorly built structures have capsized. On one occasion, a Yellow-billed Cuckoo laid two eggs in a nest of an American Robin (*Turdus migratorius*), in which the robin also laid an egg. Then a Mourning Dove (*Zenaida macrura*) added two eggs to the mixed set and incubated along with the cuckoo. Both birds were found sitting side by side on the eggs of three species.

REFERENCES: Bent 1940; D. E. Davis 1940a, 1940b, 1942; ffrench 1973; Hudson 1920; Köster 1971; Marchant 1960; Rand 1953; Skutch 1959, 1966, 1983a; Smith 1971; Vehrencamp 1977, 1978.

9.

HOATZINS

Opisthocomidae

As the gunboat *Amazonas* of Peru's inland navy wound up the muddy Río Huallaga, we passed long stretches where the high earthen banks had been undercut by the river and tall trees on the brink had fallen into the water. Among the branches of one of these trees stranded in the shallows by the shore rested a flock of large birds, who were reluctant to move as our vessel passed noisily close by them. At most they shifted to somewhat higher perches in the same prostrate tree, deliberately, one or a few at a time. Their necks were long and thin, their flight slow, loose-jointed, and clumsy; but their spreading crests of long, pointed, brown feathers gave them an aspect of alertness that their sluggishness belied. About 24 inches (61 centimeters) long, they had short, thick bills and red eyes set amid bare blue skin. Their upperparts were largely bronzy olive-brown, streaked on hindneck and mantle with buff. Their long, dark brown tails had broad whitish tips. The foreneck and breast were buffy, passing into chestnut on the more posterior underparts. After each short flight, they held their wings outstretched, displaying a beautiful patch of chestnut in the center of each. I noticed no difference between the sexes. Who that had seen pictures of this strange bird, sole member of its family and like no other in the world, or read William Beebe's classic account could fail to recognize

Hoatzins

the first living Hoatzin (*Opisthocomus hoazin*) that he met?

As we voyaged along the upper Amazon and its great tributaries or pushed in smaller boats into the backwaters, we met other flocks of Hoatzins, each consisting of from ten to about fifty birds. Their favorite resorts were among the branches of a dead tree fallen along the shore or the boughs of a living one projecting over the water; they seemed never to venture many yards inland from the bank. They voiced a variety of notes that were far from melodious. With weak, inefficient wings, they flew with reluctance and never far.

The food of Hoatzins consists almost wholly of the foliage, flowers, and fruits of certain plants that grow in shallow water and marshy places, chiefly the tall, canelike aroid *Montrichardia* and the White Mangrove *Avicennia*. To accommodate the great mass of this coarse provender that they must digest, they have developed large, strong, muscular crops, which when full so overweigh these weak-footed birds that they rest their breasts against a branch while they perch. A patch of thickened skin protects the breast from abrasion. They are said to vary their diet with a few fish and small crabs, which they capture in the mud or shallow water beneath overhanging branches.

Social at all seasons, Hoatzins build crude, flat nests of loosely interlaced twigs, apparently always amid branches at no great height above the water. Here they lay from two to four, rarely five, buffy eggs, sprinkled with brown or bluish spots, which can often be counted by peering upward through the flimsy platform of sticks. Such a nest may be attended by two to six adults of both sexes, who apparently share all the tasks of building, incubating, and rearing the young. As one replaces another on the nest, they greet with formal bows. They sit closely and may peck the nose of a visitor who peers over the rim of the nest.

After about twenty-eight days of incubation, the young hatch nearly naked, but they soon acquire a coat of reddish down. Like no other modern nestling, the newly hatched Hoatzin bears at the bend of each wing two large, sharp claws, reminiscent of the three which the adult *Archaeopteryx*, the earliest-known feathered creature, wore about 125 million years ago. Unlike *Archaeopteryx*, the Hoatzin loses its wing claws as it matures. When hungry, the nestling voices musical pipings which induce an attendant to pump up a mush of predigested foliage and open its mouth, into which the youngster inserts its head to eat. When only ten to fourteen days old, Hoatzins begin to help themselves to the foliage around their nests.

In other ways, too, Hoatzins are exceptionally precocious for altricial chicks. Long before they can fly, they venture from the nest and scramble through the surrounding tangle of branches and vines, seizing twigs with their feet and the claws on their wings, hooking the bill or the whole head over a branch for additional support. If threatened, flightless nestlings unhesitatingly jump from the platform of sticks and drop into the water from 6 to 20 feet (2 to 6 meters) below—nests are usually situated where nothing obstructs their free fall. Paddling with wings and legs, the young birds swim under water to the tangled growth of the riverbank, into which they scramble and remain until, when the danger has passed, an adult guides them back to the nest. Sometimes as many as five grown birds spread their wings to form a protecting canopy above a nestling while it climbs through the branches.

Because of the nature of their riverine habitat, it may be long before anyone makes a study of the Hoatzin's nesting comparable in thoroughness to those which we have of many other cooperative breeders. However, enough is known to suggest a relationship to the anis and Guira Cuckoos (*Guira guira*). Not only do Hoatzins build similar communal nests of sticks, but their nestlings jump out when disturbed much as infant anis do. Indeed, an adult Hoatzin looks much like an overgrown Guira Cuckoo.

The affinity of Hoatzins to anis and Guira Cuckoos, suggested by behavior and appearance, is supported by studies of their structure and egg-white proteins. Although long included among the gallinaceous birds, the Hoatzin should be classified, if not in the cuckoo family, at least close to it. However, such a strange bird seems to deserve a family all to itself.

REFERENCES: Beebe 1918; Grimmer 1962; Sibley and Ahlquist 1973.

10.

SWIFTS

Apodidae

The seventy species of swifts are distributed over most of the world, except for high latitudes, New Zealand, and oceanic islands. From 3.5 to 10 inches (9 to 25 centimeters) long, they lack bright colors but are sooty black, brown, or gray, often with prominent patches of white. They have long, tapering wings, short, weak legs, and sharp toenails for clinging to vertical surfaces. The tail feathers of some are tipped with sharp spines for additional support. Most aerial of small birds, swifts spend most of their active hours circling high in the air, their tiny-billed but broadly gaping mouths catching the flying insects that sustain them. Their unmelodious voices may be weak or piercingly shrill. With saliva from glands that become enlarged in the breeding season, they glue nests of twigs, straws, moss, or feathers to vertical cliffs, walls, or hanging palm leaves. Others construct looser nests in crannies in buildings or on rock ledges, sometimes behind waterfalls. Some species of swiftlets (*Collocalia*) make nests wholly of saliva in seaside or montane caves. Both parents incubate one to six elongated, plain white eggs and feed the young.

As spring advances northward, vast numbers of little, sooty brown Chimney Swifts (*Chaetura pelagica*) leave their winter home in western South America to accompany it, migrating by day and feeding as they go, as is the way of aerial insect eaters. In the evenings, these migratory swifts pour down into tall chimneys to sleep, often to be trapped as they emerge the next morning and ringed with numbered bands that may help trace their long journeys. Soon they spread out to breed over southern Canada and nearly all of the United States east of the Rocky Mountains. Formerly they nested in hollow trees and crannies in cliffs, but when Europeans settled in America and built houses with chimneys, the increase in the number of nest sites was probably followed by a great increase in the number of swifts. Now, in the warmer months when many chimneys are smokeless, swifts breed in them, with sticky saliva fastening to sooty vertical walls and gluing together twiglets that they break from trees with their feet as they fly by. The unlined nest, a shallowly concave shelf or bracket projecting from the wall, may be used year after year, often with repairs or additions at the outset of a new breeding season.

In its dimly lighted nest the Chimney Swift lays three to six, most commonly four or five, pure white eggs, which the sexes alternately warm during an incubation period of eighteen or nineteen days. Hatched sightless and downless, the noisy nestlings are fed with regurgitated insects by both parents. Soon they bristle with long pinfeathers, and after fourteen or fifteen days their eyes start to open. About this time the young birds begin to climb out of the nest and cling to the wall beside it, at first briefly; but when nineteen or twenty days old they spend nearly all their time on the wall, returning to the nest only when alarmed or excited and remaining only a short while. Unless disturbed, they first fly into the open air when twenty-five to thirty-two days old, but they return to roost in the chimney with their parents for about two weeks more, after which they vanish. For another week or ten days, only two swifts, apparently the parents, enter the chimney in the evenings. After this, it remains vacant.

The foregoing account of the nesting of Chimney Swifts is based largely upon observations by Althea Sherman, who early in the present century built a tower on her Iowa farm with a simulated chimney specially designed for watching its occupants. One day, while she watched a nest, three adult swifts arrived, one after another, and fed the nestlings; then all remained clinging to the wall beside the nest, leaving no doubt about their number. For at least several days, the helper continued to attend the nestlings. After the young climbed out, all six of them, plus the parents and their assistant, nine in all, could be counted by lamplight as they clung close together to the wall near the nest. One night Sherman found three adults on a nest where incubation was in progress. She commented upon the perfect harmony that prevailed among all members of the family—the young in the nest and their adult attendants. The Chimney Swift, she admiringly remarked, might be chosen as the emblem of peace.

Long before Sherman built her swifts' tower, Mary Day had watched a third attendant in a chimney in New Jersey. Many more examples of cooperation were disclosed by Ralph Dexter among swifts that nested in air shafts on the campus of Kent State University in Ohio. In eight years, he found twenty-two trios of nest attendants and six quartets. Forty individuals participated in the trios and eighteen in the quartets, some of them in consecutive years. The extra individuals, males more often than females, helped incubate, brood, and feed the young, often serving at the same nest throughout the breeding season. After the completion of a nest, two of the associates might sleep on it while one clung below or beside it, or while one slept on the nest, the other two roosted just below it. At the end of the nesting season, when the juveniles had left, trios or quartets often continued to lodge in the shaft, clinging side by side while they slept.

Some of these helpers were yearlings, who probably assisted established pairs because they could not find mates, since Chimney Swifts are capable of breeding when about one year old. Some individuals who associated with nesting pairs at the beginning of the breeding season abandoned them when they found mates and could establish nests of their own. Other helpers were old birds in their last year of life. Since sets of more than six eggs are rare, it is unlikely that two breeding females occupy the same nest; but the possibility of polyandry cannot be excluded. In any case, most Chimney Swifts nest in monogamous pairs without helpers, and occasional assistants appear to be nonbreeding individuals.

In 1950, Richard Fischer watched a pair of Chimney Swifts who occupied their nest of the preceding year and were associated with a yearling who had hatched in it. In many hours of observation, he failed to see the young swift incubate, but once after nightfall he found it on the nest beside a parent. After the nestlings hatched, the yearling fed its siblings more often than either of the parents and possibly more than both together. At night it helped brood the nestlings, sitting beside one or both of the parents. One wonders whether this young swift accompanied its parents on the long journey to South America and continued to associate with them through the months of their sojourn there. Or did it join them again at its natal spot in the spring, after a long separation? If the first of these alternatives is true, and we had a few more similar cases, we might include this long-distance migrant among the advanced cooperative breeders.

The related Ashy-tailed Swift (*Chaetura andrei*), widespread in South America and equally drab in its plumage, has similar but apparently more complicated breeding arrangements. A nesting pair is frequently joined by one to three individuals and occasionally by a larger number. In an attic at Niterói, Brazil, Helmut Sick found nine young of different ages in or near a nest, close to which nine adults roosted. Since the clutch of this swift usually consists of only three or four eggs, it appeared that more than one female had laid in the same nest. If all nine adults fed the nine young, we may have here an example of a communal nest with nonbreeding helpers, as occurs among anis. Nest helpers may occur among several species of African swifts in which males substantially outnumber females, but to my knowledge, this has not been definitely reported.

References: Dexter 1952, 1981; Fischer 1958; L. Grimes MS; Sherman 1952; Sick 1959.

11.
WOOD-HOOPOES
Phoeniculidae

The nine species of wood-hoopoes are confined to forests and savannas of Africa south of the Sahara. From 9 to 17 inches (23 to 43 centimeters) long, all are crestless birds clad in dark metallic blues, greens, and purples. Their slender bills vary from nearly straight to very downcurved, as in the scimitar-bills. Their habits are well exemplified by the best-known species, to which this chapter is devoted.

The Green Wood-hoopoe (*Phoeniculus purpureus*) ranges across the breadth of equatorial Africa and southward to the tip of the continent. About 13 inches (33 centimeters) in length, it is a long-tailed bird whose dark plumage is glossed with green and blue. White patches adorn its wings and tail. Its short legs, strong toes, and long, slender bill are red. Females are smaller than males, have shorter, less downcurved bills, and utter different notes. Newly fledged wood-hoopoes have black bills that slowly become red during the first year or more of their lives. Versatile foragers, Green Wood-hoopoes climb over trunks and branches, using their long tails as props, while their pointed bills probe crevices in bark and wood for small invertebrates. They also search among the foliage and, especially in the dry season, seek food on the ground, probing and tearing apart cow pats. When the winged broods of termites fill the air, these birds catch them in flight.

For three years, David and Sandra Ligon studied Green Wood-hoopoes intensively on the African plateau near Lake Naivasha in central Kenya. Here the birds were permanent residents in open, parklike groves of Yellow-barked Acacia (*Acacia xanthophloea*), the only tree. Their social units ranged in size from single pairs to seventeen individuals, but the majority of the groups consisted of three to nine birds. Each group, large or small, contained a single, dominant breeding pair, composed of the oldest male and female, usually with one or more nonbreeding helpers who were often, but by no means always, their offspring or at least closely related to them.

As long as both members of a reproductively successful pair remained alive, the group tended to be a nearly closed, stable association; but the death of one of the dominant adults was followed by uprootings and reorganization. Wood-hoopoes who moved from one flock to another usually did so in pairs or trios of the same sex, either males or females who were often siblings, and typically they migrated no farther than an adjacent territory where a vacancy had occurred or one that was weakly defended. The oldest invading male might be-

Green Wood-hoopoe

come the breeder in the reorganized flock, while his companions remained as auxiliaries, perhaps to rise to breeding status after his demise. Flock members frequently preened one another. Sometimes the dominant male, after repelling a subordinate male from the breeding female, groomed the subordinate, as though to make amends for his harshness.

Each group tenaciously defends its territory against incursions by neighboring groups. In these encounters they display a behavior which is rare, if not unique, among birds, called flag waving. As they advance toward the invaders, one or two birds, usually including at least one of the breeding pair, pick up a cluster of lichens, a piece of bark, or a wad of spiderweb and frass. Waving these ensigns vigorously back and forth in the tips of their long, vivid bills, the flag bearers lead the attack, often venturing to within a few feet of the encroaching wood-hoopoes and thrusting the banners toward them, as though in defiance. Sometimes a flag is passed from one of the defenders to another, perhaps to intensify their determination to present a united front to their rivals. While they wave flags in territorial encounters, wood-hoopoes also display another peculiar behavior called the rally. Clumping closely together, group members sway their bodies from side to side and toss their tails up and down "in a ridiculously quaint fashion," while all together they voice a chorus of loud, excited, chattering notes. This rally display also greets group members who rejoin their companions after an absence. Flag waving is often witnessed while wood-hoopoes mob hawks or other predators or confront birds of other kinds who compete for the cavities in which they sleep and nest.

At all seasons, Green Wood-hoopoes lodge, singly or up to seven together, in cavities in trees, which may be old woodpecker holes or hollows resulting from decay. Pugnacious Blue-eared Glossy Starlings (*Lamprotornis chalybaeus*), abundant Acacia Rats (*Thallomys paedulcus*), aggressive African Honeybees (*Apis mellifera*), and other creatures compete for these hollows with the wood-hoopoes, who sometimes can find no better shelter than loose flakes of bark. Apparently, wood-hoopoes must sleep in cavities to maintain their body temperature during the cool nights of the plateau, 6,400 feet (1,950 meters) above sea level. When disturbed in their dormitories in the night, wood-hoopoes turn their tails toward the entrance and—from their exposed oil glands—expel drops of fetid oil that may repel predators or perhaps the Acacia Rats that covet their holes. Nevertheless, nocturnal predators appear to be a principal source of wood-hoopoe mortality.

Usually in June, four to six weeks after the advent of the "long rains," when the lepidopterous larvae indispensable for successful breeding become abundant, wood-hoopoes begin to nest. On the unlined floor of a nest cavity from 3 to 72 feet (1 to 22 meters) above the ground, the dominant female of a group lays two to four plain white eggs that may differ conspicuously in size. During an incubation period of approximately eighteen days, she warms these eggs unaided, while she is so well nourished by her mate and other group members that she need find little or nothing for herself. At twenty-one carefully watched nests, the number of attendant helpers varied from one to seven, and the percentage of feedings attributed to them ranged from 20 to 81. The breeding male contributed the rest of his mate's food; the more assistants he had, the less he brought. As group members approached the nest cavity, the incubating female solicited food with a rapid, continuous twittering. If her attendants were negligent, she might leave the hole and pursue them while constantly emitting this call, which usually evoked the desired response.

Hatching failure, not attributed to egg loss, was surprisingly high. At least one egg failed to hatch in nine of ten nests that were carefully checked. Although four was the most common clutch size, only once among thirty-seven broods were four fledglings counted. The young wood-hoopoes hatch with long, loose down, the callose heel pads usual on nestlings that grow up on a hard, unpadded floor, and conspicuous pale swellings, reminiscent of those of nestling woodpeckers, at the corners of their mouths. During approximately the first half of their thirty days in the nest, their mother stays almost continuously with them, while she and they are fed by the father and by helpers who may contribute up to 83 percent of all feeding visits to a nest, thereby substantially reducing the work of the parents. At an exceptionally late nest in January, the nestlings were fed by nine adults plus three juveniles fledged in July, a total of twelve attendants. At first, much of the food is delivered to the mother for transfer to the nestlings, but as the young birds grow older the attendants prefer to feed them directly. The principal food of the nestlings, as of their mother before they hatch, is the larvae of butterflies, moths, and beetles.

When the nestlings were old enough to perch at the entrance to the nest cavity, all flock members groomed them and gave vocal rallies in front of them, thereby promoting the intimacy that would bind group members together. Certain helpers, whom the Ligons called aunties, were outstandingly attentive to nestlings. While their mother incubated, future aunties fed her no less regularly than other group members did, without trying to displace her. After the young hatched, instead of passing food to the twittering mother for transfer to them, as other attendants did, the auntie tried to avoid her and feed them herself. She might enter the nest and remain as long as ten minutes, twittering and importuning other group members for food, which she passed to the nestlings. Some aunties spent much time inside the nest while the mother was present. Most mothers did not oppose the aunties, who were generally known or presumed to be their daughters; but at two nests, where the would-be aunties were unrelated to the nestlings, they were hindered or mildly threatened by the mothers. Aunties guarded nests and protected them from intruding birds of other species, including starlings and hornbills.

After the young wood-hoopoes fledged, all group members continued to feed and protect them from predators. Late in the afternoon, the old birds, sometimes as many as nine together, cooperated to "tuck in" the youngsters or put them to bed. First, the older birds bunched together in a tree and directed rally calls to the fledglings. After the young joined them, the adults flew by stages to the roost tree, repeating rally calls along the way, guiding the inexperienced birds to their dormitory. When, with much excitement, the family reached the chosen cavity, an adult usually entered first and the young followed. Still hungry, they begged for food and were actively fed inside the hole. After an interval of foraging for themselves and quietly preening, some of the adults might join the fledglings for the night. For about two weeks after their emergence from the nest, the young wood-hoopoes were led to bed in this fashion, not always in the same cavity. Commonly they alternated between two or three holes, probably for greater safety from predators.

The greater the number of auxiliaries, the less the male parent worked to provision his offspring. However, the Ligons found no consistent relationship between group size and the number of young fledged per nest or the number surviving until the year's end. Over three years, groups of more than five birds produced an annual average of 2 juveniles per group, whereas smaller groups produced an average of only 1.3 young per group. Although in two of the three years the larger groups produced more young per group, in two of these same three years the smaller groups reared more surviving young per flock member. The maintenance of a group appeared to be more dependent upon the survival of adults than upon the number of eggs laid or young hatched. Differences in reproductive success were not consistently related to the quality of the wood-hoopoes' territories.

Throughout the Green Wood-hoopoes' great range, from Kenya in the east to Ghana in the west and southward to Cape Province, nest helpers have been found among them. Helpers also occur at least occasionally in the related Hoopoe (*Upupa epops*), but no thorough study of this very widespread bird is available.

References: L. G. Grimes 1976; Ligon and Ligon 1979.

12.

TODIES

Todidae

Five species of charming little birds constitute the tody family, which is confined to the Greater Antilles. Two species are found on Hispaniola, one on Cuba, one on Jamaica, and one on Puerto Rico. Slightly over 4 inches (10 centimeters) long, all are quite similar in appearance, with broad, flat bills, bright green upperparts, wings, and tails, red throats and lower mandibles, largely white underparts, and yellow under tail coverts. The sexes are always alike. Most ornate is the Cuban Tody (*Todus multicolor*), which has a blue patch on each side of its neck and pink flanks. The flanks of the Puerto Rican Tody (*T. mexicanus*) are yellow instead of pink. The todies' closest relatives are the much larger motmots of mainland tropical America.

Best known is the Puerto Rican Tody, which Angela Kepler studied intensively for three years. A bird of great ecological tolerance, it is widely distributed over the island, from semidesert scrub at low altitudes to dripping rain forests near the mountaintops. It thrives in coffee plantations, bushy pastures, and second-growth thickets. Its diet consists largely of insects, which on short flights it plucks from the upper or undersides of leaves, from trunks and branches, or from the air, usually at no great height. To support its rapid metabolism, this tody captures small insects at the rate of one or two per minute. It has a variety of notes, ranging from a short beep to a trill; and with its slightly attenuated outer wing feathers it produces a rattling, whirring sound like that made by passing a finger rapidly over the teeth of a comb.

Todies never flock but live at all seasons in monogamous pairs on defended territories which, if all goes well, they occupy from year to year. The male feeds his mate. They nest in burrows which the two sexes, taking turns, dig in preferably vertical banks, often under sheltering roots or rocks, in cut banks beside a road or trail, and, in drier regions where the danger of flooding is slight, in streamside banks. About 12 inches (30 centimeters) long, rarely twice that length, the tunnels may be straight but often bend sharply to the right or left. At the inner end each expands into a low chamber where one to four, usually two or three, glossy white, extremely fragile eggs are laid upon the bare floor.

Although eggs are laid between March and July, excavation of the burrows may start as early as the preceding September. These earliest burrows appear never to be used for eggs, but three or four weeks commonly elapse between the completion of excavation and laying. In starting to dig long before they will nest, todies resemble certain other burrow nesters, including motmots. Todies do not sleep in their burrows in the long interval before laying begins.

Both parents develop bare brood patches and take turns at incubation, but to the female falls the larger share of the task. Like other very small insectivores, todies sit inconstantly, taking sessions that average only thirteen minutes and rarely continuing in the burrow for as long as an hour by day. At four of twenty-three closely watched nests, helpers participated in incubation but their contribution was slight. After an incubation period of about twenty-one days, the nestlings hatch, wholly naked, with tightly closed eyes, and on each heel a thick pad of swollen skin rough with numerous tubercles, which protects the heel joint as the young todies stand and shuffle around on the burrow's uncarpeted floor. After developing very long pinfeathers, the nestlings become well clothed with green plumage at the age of about nineteen days. Like many other nonpasserine burrow nesters, the parents neglect sanitation, and the chamber becomes foul with excreta and the remains of insects.

Helpers, who were not present during the excavation of the burrows and who shared incubation at less than one-fifth of the nests, became more numerous at burrows with young, where they might appear as late as the sixteenth day of the nestling period. One or two of these auxiliaries, of both sexes but predominantly males, were active at about half of the nests. At first repelled by the parent todies, after a few days they were accepted and contributed substantially to the nourishment of the young. At a burrow watched for thirteen hours, the presumed parents made 62.5 percent of the feeding visits, and their two assistants made 37.5 percent. With the increased food that the helpers provided, the nestlings developed more rapidly and left the burrows at an average age of seventeen days, two days sooner than at burrows without auxiliaries. At nests with helpers the average clutch size was 2.9, whereas at nests attended by the parents alone it was 2.2. At nests with helpers an average of 2.6 young were fledged, but only 2 were fledged at unaided nests. However, the average number of fledglings per adult was only 0.72 at nests with three or four attendants, but it was 1 at burrows where only the parents fed the nestlings.

The relationship of the helpers to the parents whom they assisted was puzzling. Evidently they were not, as in many other species of cooperative breeders, yearling todies who had remained with their parents over the nonbreeding period, when each territory contains only two grown todies. They were not present during the excavation of the burrows, and when they appeared late in the incubation period or after the nestlings hatched, they were at first chased by the resident pair. Since todies rarely lay twice in a season, some of the helpers may have been breeding adults who, after the loss of their own nests, turned their attention to the nests of more fortunate neighbors, as in Long-tailed Tits (*Aegithalos caudatus*); others could have been individuals who had failed to secure a mate and a territory.

Even more puzzling was the fact that nests with auxiliaries contained substantially more eggs than nests without helpers. Since the assistants were apparently not present when the eggs were laid, how did the resident female know that she would receive aid that would enable her to rear a larger family? Could some of these helpers have added eggs to the clutches, undetected by the careful observer? In other species of birds, older females frequently lay more eggs than those breeding for the first time; could it be that it was chiefly the older todies who attracted assistants? Possibly todies who lacked young of their own remembered, and returned to, the territories where they themselves had been reared and helped their own parents rear their younger siblings. If this were true, the todies' behavior would conform to the theory of kin selection. Despite an exceptionally prolonged and painstaking study of cooperative breeding, a few questions remain unanswered.

After leaving the nest when nineteen or twenty days old, Puerto Rican Todies receive most of their food from adults until they are about twenty-seven days old. From this age onward they catch an increasing proportion of the insects that they consume, until at about forty-three days, or twenty-four days after leaving the burrow, they are fully independent. Soon thereafter, they vanish from the parental territory.

REFERENCE: Kepler 1977.

13.

KINGFISHERS

Alcedinidae

Most of the eighty-six species of kingfishers live in the eastern hemisphere, where they are most numerous in southeastern Asia, Indonesia, and islands of the southwestern Pacific. Only six species are found in the New World, from Alaska to Tierra del Fuego. In length kingfishers range from 4 to 18 inches (10 to 46 centimeters). Their bills, especially those of the fishing species, are long, stout, and sharp-pointed. The New World kingfishers, clad in deep green or gray, with white and chestnut underparts, lack the diversity and brilliance of Old World species, many of which are splendid in red, orange, yellow, green, or blue, often with vividly red bills. Some have elongated tails with spatulate tips. All the American kingfishers are primarily what their name implies; but many of the Old World species are forest dwellers, who may live far from open water, nourishing themselves with insects, other terrestrial invertebrates, and even small vertebrates. The aquatic kingfishers excavate burrows in streamside and other earthen banks; the forest dwellers nest in holes in trees or termitaries. All lay from one to ten white eggs in unlined chambers that they usually fail to keep clean. Both sexes incubate and attend the young, who remain in the nest for three to five weeks.

All the New World kingfishers that have been studied nest in solitary, monogamous pairs, but some of the Old

World species breed cooperatively. Among them is the Laughing Kookaburra (*Dacelo gigas*), one of the largest and heaviest of kingfishers, common in woodlands of eastern Australia, where it is native, and southwestern Australia, where it has been introduced. Largely dark brown above and white lightly barred with brown below, with rufous on the tail, blue on the rump of adult males, and a little blue on the wing coverts of both sexes, it is one of the least colorful members of its family. Its bill is relatively short and very stout. Its famous laugh, probably the most widely familiar of Australian bird sounds, is preceded by a low chuckle given by one member of a group. As others join in, the chorus swells to a crescendo of *ha-ha-ha* laughter, then subsides to restrained chuckles. Neighboring groups often respond with similar boisterous outcries, especially at dawn and at nightfall, which has earned for this kingfisher the sobriquet "bushman's clock." Kookaburras occasionally plunder birds' nests or pilfer goldfish from ornamental ponds; but they subsist mainly upon insects and devour many young snakes, so that, as Ian Rowley said, "on balance they do more good than harm."

The social and breeding habits of Laughing Kookaburras were studied for two years by Veronica Parry in a wooded residential district at Belgrave, Victoria, on the western edge of Sherbrooke Forest in the Dandenong Ranges. There the kingfishers lived throughout the year in simple pairs or in groups of three to six

Laughing Kookaburra

individuals, each including a single breeding pair with one to three or, rarely, four helpers, the breeding pair's progeny of previous seasons, who comprised about a third of the adult population. Although physically able to reproduce when slightly less than twelve months old, these auxiliaries may remain subordinate and refrain from breeding for one to three or more years. They are of both sexes, for among kookaburras, in contrast to many other cooperative breeders, males and females occur in approximately equal numbers.

In each generally peaceful group of kookaburras, a hierarchy developed and was maintained by sparring matches that resembled Indian arm wrestling. Two birds would grasp bills and twist and turn their heads and bodies until one was either thrown from the perch or admitted defeat and retreated. These tests of strength might last for thirty seconds to fifteen minutes, while the contestants used their wings for balance and to beat one another. Perching alone amid dense foliage, young fledglings spent hours practicing for these contests by pecking at and twisting leaves and branches. When about three months old and nearly independent, nest mates of the same sex engaged in these wrestling bouts. The outcome of all such engagements was highly predictable: nobody was hurt, but the breeding male always prevailed over all others; the oldest male helper defeated the next oldest, who in turn could throw a fledgling from a perch. The hierarchy among the males was established in this order. Females had a separate order, with the mother dominant over helpers of the same sex. After feeding fledglings, parents and helpers frequently wrestled with them. If the young birds challenged their elders, they always lost. Thereby they learned their rank in the group.

Each group of kingfishers maintained throughout the year a territory that supplied all its needs. In area the territories ranged from about 5 to 25 acres (2 to 10 hectares), but those of the same group fluctuated in size from year to year: if a group increased its membership by successful breeding, it enlarged its domain at the expense of less fortunate neighbors, whose holdings shrank if their groups became smaller. After annual adjustments, a territory included about 3.2 acres (1.3 hectares) for each of its occupants. This area appeared to depend less upon its productivity of food (although this set a lower limit to its size) than upon the number of birds able to defend it by visual and vocal displays between opposing groups at definite points along their common boundary. Only once, when a fledgling wandered into the territory of a neighbor, who tried to feed the young interloper before its father crossed the boundary to retrieve it, did Veronica Parry see two kookaburras clash physically.

In Victoria, the kookaburra's breeding season extends from September to late November or December. The courting male feeds his mate. One female was so badly injured that her consort had to hunt food and feed her continuously. Although permanently disabled and dependent upon him for survival, she tried to breed and laid one egg before she died, four months after she was wounded. Kookaburras occasionally excavate nesting cavities in termitaries and have nested in such strange places as bales of hay, holes in city buildings, and the ash pile of an operating forge. In Parry's study area, the single breeding pair in each group nested low or high in a tree, in a natural hollow which occasionally they remodeled with a little carving. A doorway at least 5 inches (13 centimeters) in diameter opened directly into a domed chamber usually about 12 inches (30 centimeters) wide by 9 inches (23 centimeters) high. Unlike the cavities of most hole-nesting birds, it did not extend below the opening. Not more than 20 or 30 feet (6 to 9 meters) in front of each nest was a perch where the birds always alighted before entering or after leaving the hole. This nest perch was a center of activity while nesting continued.

On the chamber's unlined floor, about level with the doorsill, the female kookaburra laid from one to five, usually two to four, white eggs. The mean clutch size of females of simple pairs was 2.5, while that of females with helpers was 3; probably the latter were older birds. Incubation, which began soon after the first egg was laid, was performed by all group members. Like the parents, the nonbreeding helpers of both sexes developed bare, swollen brood patches well supplied with blood vessels, so that they effectively warmed the eggs. At different nests, one or more helpers incubated for 14 to 25 percent of the observation periods; at one of these nests, their contribution to incubation exceeded that of a parent.

The helpers would have sat longer had they been permitted to do so by the primary incubator, who in different groups was either the breeding male or the breeding female. Usually a bird coming for its turn on the eggs alighted on the nest perch and called a soft *kooaa* followed by a *chuckle* or *chuck*, which brought the sitting bird to the doorway. After responding with similar notes, it flew out and the newcomer entered. The primary incubator would not always leave when another arrived and signified its intention to incubate, in which case the latter flew away. But when the primary incubator wished to enter and the bird already present refused to leave, it would go sit beside the other. The two might remain side by side for up to five minutes before the bird who was there first flew away. The length of the sessions of any individual and the sequence in which the several collaborators sat on different days were most variable. Both family groups and simple pairs left their eggs unwarmed for about one-quarter of the daytime. At night only the breeding female was found in the nest.

After an incubation period of twenty-two to twenty-five days, the nestlings hatched blind and quite naked, as in other kingfishers. Unlike kingfishers that nest in burrows underground and permit them to become shockingly foul with excreta and regurgitated indigestible food, the young kookaburras kept their hole clean by turning around, backing up, and squirting their liquid excrement through the doorway—which is probably the reason why the nest cavity did not extend below the doorsill.

The nestlings were brooded and fed by all members of their group, at some nests the father, at others the mother, and at still others the auxiliaries contributing the major share of food. Even a helper with a badly injured leg brought 14 percent of the meals delivered at a nest. Often an attendant arriving with food gave it to a brooding adult for delivery to a nestling. Nestlings were fed with about equal frequency by simple pairs and by larger groups. They grew at the same rate and when a month old were almost as heavy as the average adult. They had outgrown their infantile ugliness and were fully feathered, but they lingered in the nest a few days more, gaining strength to fly. They were well guarded by attendants who darted close to Parry's head when she climbed a rope ladder to examine them, once striking the back of her head so hard that thereafter she wore a safety helmet.

When, at ages ranging from thirty-three to thirty-nine days and averaging thirty-six days, the young kookaburras were ready to leave their nest cavity, they advanced to the doorway and sat there from two to four hours, stretching but not flapping their wings, while the adults perched in view, repeating feeding calls and sometimes eating the food that they had brought for the young. Finally, becoming hungry, a fledgling launched forth and flew about 20 yards (18 meters) to alight on the ground. Here it was joined by the adults of its family, who, as though celebrating its graduation from the nest, sang in chorus until the fledgling rose to a low perch, where it rested quietly for the remainder of the day. Unlike the young of many cooperative breeders, fledgling kookaburras were never led to sleep in the nest or another dormitory but passed the night amid concealing vegetation, where the adults joined them at daybreak. The day after their departure, they were led away to some spot among dense foliage. Soon they could fly up and join the adults on the roost where all slept. For eight weeks after they fledged, the young kookaburras were dependent upon the adults for food, which they clamored for with a ceaseless harsh, grating din. Then, if their parents hatched a second brood, they might help feed their younger siblings. Although far from expert, one juvenile contributed 10 percent of the meals brought to them.

Simple pairs and family groups nested with almost equal success, rearing fledglings from about half of their eggs. However, nine simple pairs produced only 1.2 fledglings per pair, whereas with slightly larger clutches and second broods, ten family groups raised 2.3 fledglings per family. After the young fledged, the difference in the two categories was greater. By the beginning of the next breeding season, all the offspring of pairs had died, but six of ten of those in families still lived. Two factors contributed to their greater survival. In the first place, parents without assistance unavoidably left their fledglings alone while seeking food for them; but young in family groups were rarely without a guardian, who might protect them from predators while they spent much time on the ground learning to forage. Moreover, in family groups the young could be fed to a more advanced age. This was because parents often started to molt before their offspring became fully self-supporting, and while heavily molting they tended to become solitary and to rebuff their begging progeny, who might succumb in the absence of other attendants. In a family group, the helpers, who molted earlier than the parents and sometimes finished before the molt of the breeding birds had advanced far, could take full care of the fledglings until they became self-supporting.

When, instead of molting, the parents attempted to rear a second brood, the helpers took full charge of the young kookaburras, leaving the breeding pair free to nest again about two months earlier than they otherwise could have done and to give undivided attention to their new undertaking. Only family groups nested successfully twice in a season; unassisted pairs did not renest. As in Florida Scrub Jays (*Aphelocoma coerulescens*) and some other cooperative breeders, the helpers demonstrated their value most convincingly during the perilous interval between the fledglings' departure from the nest and their attainment of independence.

Another cooperatively breeding member of this family is the black-and-white Pied Kingfisher (*Ceryle rudis*), widespread in Africa south of the Sahara and in southern Asia. It is said to be the most "pelagic" member of the family, sometimes plunge-diving from the air into the open water of a lake nearly 2 miles (3 kilometers) from shore. Unlike most kingfishers, the Pied can catch two fish in one dive and can swallow at least small ones in flight, although it follows the usual practice of carrying larger prey to a perch and beating it before swallowing it. It varies its fare with termites, aquatic insects, crustaceans, and an occasional grasshopper.

Pied Kingfishers are the most colonial

Pied Kingfishers

members of their family. On the northern shore of Lake Victoria in Uganda, Robert John Douthwaite studied a colony for two years. Here the birds dug nesting burrows in borrow pits and banks of ditches within six-tenths of a mile (1 kilometer) of the lake. A male and a female, digging alternately for an average of twenty-six days, finished tunnels whose length of 31 to 98 inches (79 to 250 centimeters) varied inversely with the hardness of the soil. During the weeks of digging, groups of up to six or eight kingfishers, mostly males, displayed on open ground with upright posture, curiously elevated wings, and open bills. Also during excavation, the male began to feed his partner, and immediately before and during the laying period he provided most of her food; she seldom flew except to take it from him. A few days after completion of the burrow, she began to lay her three to six (most often five) porcelain white eggs, depositing them on consecutive days on the bare floor of the chamber at the end of the tunnel. Both sexes incubated, the female more than the male, who spent much time on the ground in front of the burrow.

When another kingfisher approached the burrow, the guarding male usually threatened the intruder, who gave an appeasement call and retreated. Sometimes the visitor, always a male, advanced with a fish held head outward in his bill. The guarding male usually seized the fish, starting a struggle which ended with his taking the fish from the newcomer and eating it. The only occasion when the intruding male was seen to feed a mated female occurred immediately after a changeover at the nest, when her partner had disappeared inside. This time the fish was relinquished without the usual struggle. Throughout the year males, who in large colonies might be 2.4 times as numerous as females, fed one another, sometimes at a distance from any colony; but such feeding was most frequent outside burrows containing incubating females.

Because incubation began before the clutch was complete, the first nestling might escape from its egg as much as three days before the last, which hatched after an incubation period of eighteen days. Both parents fed the nestlings, assisted at some burrows by up to three or four helpers, who were always males, some of them older progeny of the pair whom they assisted to feed their siblings, others unrelated individuals. Of ninety-six feeding visits at several burrows, sixty-nine (72 percent) were by males. When, at the age of twenty-four or twenty-five days, the young Pied Kingfishers left the nest, at least some helpers continued to feed them. Fledglings were usually fed in the air after a chase, but they returned to a perch to swallow their food. Three days after leaving the burrow, they were shaking and battering their meal, and within two weeks they were diving into the water, probably in pursuit of fish. For at least a month after fledging, they continued to receive food from adults.

Pied Kingfishers were also studied in East Africa by Heinz-Ulrich Reyer, who found them nesting in colonies of twenty to one hundred or more birds, in burrows separated from their nearest neighbor by an average distance of 17 feet (5.2 meters) at Lake Victoria and 5.25 feet (1.6 meters) at Lake Naivasha, with some as close together as 20 inches (51 centimeters). In these colonies, adult males outnumbered females by 1.8 to 1, although among nestlings and fledglings the sexes were represented equally. By spending more time in the nests by day and performing all the nocturnal incubation and brooding, females suffered higher losses from predation and the collapse or flooding of burrows than males did, and their greater tendency to disperse from their natal colonies also exposed them to greater risks.

Reyer distinguished two kinds of helpers. Nearly a third of the pairs in both colonies had primary helpers, yearling sons of at least one member of the pair that they assisted, who returned with their parents to their nest site of the preceding year. While the parents excavated burrows, courted, and laid eggs, these helpers fed the breeding male directly, the female not only directly but also indirectly by giving the male food that he passed to her. They did not assist in digging, but they helped the pair drive rivals from the nest site. After eggs were laid, these primary helpers joined the parents in defending the nest from predators, and in due time they fed the young kingfishers. They were not seen to incubate or brood.

Secondary helpers, also yearling males, appeared not to be closely related to the pairs they joined. During digging, laying, and incubation they remained aloof from the burrows, appearing three or four days after the eggs hatched, probably attracted by the nestlings' loud, begging cries. Fish in beak, these would-be helpers flew through the colony, alighted near various nesting burrows, and waited. After a while, they repeatedly approached an entrance and tried to feed the female parent. In both colonies, they were at first repelled, especially by the male, just as they were at Douthwaite's colony. At Lake Victoria, a persistent volunteer was accepted after trying from two to four days to be helpful; then he fed and protected the brood. At Lake Naivasha, on the contrary, none of nine yearling males who tried to attend nestlings was permitted to approach them. Here only primary helpers were tolerated, probably because of persisting ties with their parents and because, since cooperatively breeding birds avoid incest, they offered no threat to the father's sole paternity of the latest brood.

Not only unattached yearling males became secondary helpers; a paired male who could not finish, or lost, his burrow might serve at the nest of a more fortunate neighbor. In three such occurrences, the females of these auxiliaries remained attached to their mates but did not join them in helping. Likewise, a potential or actual primary helper might become a secondary helper. After his mother was killed by a nest predator, a yearling male and his father brought fish to the entrances of various burrows, at one of which the son was accepted as a coworker, while the father was chased away. Another yearling became a helper after his father and stepmother abandoned an uncompleted burrow and disappeared from the colony.

The acceptance of a secondary helper at Lake Victoria evidently depended upon the nesting pair's need of assistants. The only pairs who continued for a long while to reject proffered aid were two that hatched only two nestlings, instead of the usual four or five. When the nestlings of these two still unassisted pairs were twelve days old, Reyer added two young of similar age to each brood, with the result that within three days both pairs had accepted a secondary helper. Of forty-one pairs at Lake Victoria, only fourteen remained unassisted; of the others, nineteen had one helper, seven had two, and one had three, including helpers of both categories. At Lake Naivasha, five of eighteen pairs each had one primary helper.

At both lakes, pairs with and without helpers laid the same number of eggs, an average of five, and hatched nearly all of them; but the effect of helpers on the survival of nestlings differed strikingly in the two colonies. At Lake Naivasha, unhelped pairs raised 80 percent of the young that they hatched, twice as many as their counterparts at Lake Victoria, where many nestlings of unassisted pairs starved. The single helper at four of the Naivasha burrows increased only slightly the yield of fledglings per nest. At Lake Victoria, however, the contributions of a single auxiliary doubled the production of fledglings, while parents with two helpers raised every one of the young hatched in six nests, an average of 4.7 fledglings per brood. At Lake Naivasha the helper did not increase the productivity of nests per adult attendant, but at Lake Victoria one or two helpers increased by one-third the number of fledglings per adult. At Lake Naivasha, nestlings were fed at the same rate whether their parents had or lacked helpers, but at Lake Victoria the auxiliaries almost doubled the rate of feeding.

Why these great differences in willingness to accept unrelated helpers and their effect upon reproductive output between two colonies of the same species in the same region? To answer these questions, Reyer investigated the conditions in which the birds fished. At Lake Naivasha, the prevailing wind blew from the land, leaving a zone of calm, relatively clear inshore water, protected by a wall of tall papyrus, in which the birds caught fish on more than three-quarters of their dives, averaging one every 5.9 minutes. At Victoria, the wind swept across the broad lake, raising high waves that roiled the inshore water with sand and mud, so that the kingfishers had to fly twice as far to their fishing grounds as at Lake Naivasha. To catch a fish, the Victoria kingfishers spent an average of thirteen minutes flying and hovering over the lake, and only one-quarter of their dives were fruitful. Moreover, the fish most frequently brought to their young, *Engraulicypris argenteus*, were less nutritious than the fish at Lake Naivasha, so that the attendants had to bring more of them to satisfy their nestlings. In addition to these disadvantages, the frequent presence of people kept the birds away from their burrows. At Lake Simbi, the situation was even less favorable for the parent kingfishers, whose nests were about 1 mile (1.5 to 2 kilometers) from their fishing grounds—three or four times as far as at Lake Victoria. Here parents recruited three or more helpers, which they could do because the excess of males over females was apparently much greater than at the other lakes.

Reyer's study is important because it demonstrates, more clearly than any other I have seen, that the number of helpers per pair is an *adaptation to,* not a consequence of, the quality of the environment. Moreover, since territories were not defended on the open waters of the lakes and all Pied Kingfishers were free to fish in them, Reyer proved that these kingfishers do not become helpers because their chances of survival are slight if they leave the parental domain; but they are so eager to participate in parental care even at the nests of unrelated pairs that despite repeated rebuffs, they persist in offering their services until they are accepted. These facts must be remembered when we theorize about the origins of cooperative breeding.

The Striped Kingfisher (*Halcyon chelicuti*), widespread in tropical and southern Africa, nests in holes in trees. At about one-fifth of its nests, Reyer found a male helper who participated in incubating and feeding the young.

References: Douthwaite 1978; Fry 1972, 1980; Parry 1971, 1973; Reyer 1980a, 1980b; Rowley 1975.

14.

BEE-EATERS

Meropidae

The twenty-four species of bee-eaters are spread widely over the temperate and tropical zones of the Old World, from Europe to China, Indonesia, Australia, and Africa. From 6 to 14 inches (15 to 36 centimeters) in length, they have long, slender, slightly downcurved, sharp-pointed bills and short legs and are gorgeously attired in the most varied patterns of green, blue, violet, red, yellow, and brown, with little difference between the sexes. Many have long, attenuated central tail feathers that project well beyond the others, adding to the birds' grace as in swift flight they catch insects in the air. These include many bees, wasps, and other stinging hymenoptera, as well as beetles and other insects. Although the forest-dwelling species tend to be solitary, many bee-eaters that inhabit savannas and other open country are highly social. When not sleeping in their burrows, they roost in trees, often lined up in closest contact, sometimes with a few on the backs of their companions. Bee-eaters' nearest counterparts in the tropics of the New World are the unrelated long-billed jacamars, which, however, do not flock and include more butterflies and dragonflies in their diets.

One of the most thoroughly studied species is the Red-throated Bee-eater (*Merops bulocki*), widely distributed across tropical Africa north of the equator. Both sexes are largely green above, with a black facial band, buff collar and underparts, and black crissum. Gregarious at all seasons, they nest in compact colonies of usually twenty to thirty pairs, which dig yard-long burrows in high earthen banks, often near streams. At daily intervals, a female lays from one to four (average 2.6) unmarked white eggs on the bare ground of the expanded chamber at the tunnel's inner end. A pair raises no more than one brood a year.

In a wooded savanna in northern Nigeria, C. H. Fry prepared a colony for study by blocking, at night, the mouths of burrows in which bee-eaters slept. In the morning he removed the blockade, caught the birds in a net as they emerged, banded them with colored and numbered rings, and released them. To inspect the nests, he dug deep pits at the top of the bank that contacted the inner ends of the burrows, then closed each of them with a stone, a plaster of paris seal, and a halved tin with a fitting lid. After each inspection, he carefully replaced the seal. This work was done in the interval between the excavation of the burrows in November, at the end of the rainy season while the ground was still soft, and the birds' return to breed in January, when the dry lateritic earth had become extremely hard.

Red-throated Bee-eaters normally mate for life. About two-thirds of the ninety-six nests in Fry's study area were occupied by pairs without helpers. The remaining third were attended by larger groups, including twenty-three trios, three quartets, and a quintet consisting of a mated pair with three auxiliaries. The mated male and female took rather equal parts in digging the burrow, incubating the eggs, and feeding the young before and after they flew. The helpers shared all these tasks but usually did less than the parents themselves.

Populations of Red-throated Bee-eaters contain about three males to two females. Most individuals start to nest when about one year old, but a third of them defer breeding until they are two years old. Accordingly, most of the helpers are yearling males; one was two years old; another helped the same pair in two consecutive years. A yearling, whose mate disappeared two days before egg laying in his colony began, joined a neighboring pair and assisted at their nest at least until their brood fledged. A breeding unit, consisting of a mated pair and their helper(s), forms two or three months before nesting begins and is usually stable. All its members feed, rest, and sleep in their burrow in close association.

If a nest fails, the adults who attended it may migrate together to another colony. However, changes in the composition of a breeding unit do occur. Even as late as the start of incubation, a bee-eater may join or abandon the breeding unit. Since never more than one clutch was laid in a nest, it was clear that the extra females were simply helpers, never breeders. With the males, the situation was more complicated. When two males attended a nest, one demonstrated his dominance by feeding his mate, excavating, incubating, and feeding the young more than his associate did; but the other also fed the breeding female and attempted to copulate with her, so that he might have been the father of some of the young that he fed.

Thirty-six unassisted pairs produced eighty-two fledged young, and ten multiple units (nine trios and one quartet) raised twenty-seven fledglings. The multiple units raised 2.7 progeny per nest; the unassisted pairs, 2.3. However, for each adult of the unaided pairs the number of fledglings was 1.1, and for each adult of the trios and quartet it was only 0.9. Thus multiple units produced more young per nest but fewer per adult than unassisted pairs, but the difference was slight and not statistically significant.

After they fly from the burrow, young bee-eaters are fed not only by the parents and their helpers but, apparently, at least occasionally by other adults. For four to six weeks after fledging they continue to solicit food, which at first consists of a variety of insects, including many nonvenomous hymenoptera, and gradually changes to a diet of 80 percent hymenoptera, a quarter of which are stinging honeybees. In the weeks after fledging, the family of parents, helpers, and juveniles remains intact. They roost together, usually in trees rather than in burrows, and forage apart from other families along watercourses or in wooded savannas. This association persists long after the juveniles are fully self-supporting; but after six or seven months, toward the end of the rainy season, family parties seem to disintegrate as the bee-eaters forage singly. When another breeding cycle begins with the excavation of new burrows or the refurbishing of old ones, new pairs and multiple groups form as new recruits enter the breeding community. If two or more young bee-eaters of the same brood fail to find mates at this season, they appear to sever bonds with their parents but may together join another mated pair as helpers; at two nests the helpers were siblings. At other nests the single auxiliary was last year's offspring of the same pair of adults, a situation that appears to be frequent among bee-eaters, as among other cooperative breeders.

At Lake Nakuru National Park in Kenya, Stephen Emlen and his associates studied White-fronted Bee-eaters (*Merops bullockoides*), who nest in burrows 3 to 6 feet (1 to 2 meters) long densely crowded in earthen cliffs. Here the birds live in extended families or clans that comprise up to sixteen individuals of three or occasionally four overlapping generations, including two or three simultaneously nesting pairs. Throughout their lives, clan members preserve close bonds. At all seasons, they sleep together in their burrows. Singly or two or more together, they

White-fronted Bee-eaters

visit each other's burrows unopposed, although they may be repulsed from the tunnels of other clans, and they greet one another almost daily. They cooperate in digging burrows, feeding breeding females, and incubating, feeding, and protecting the young. Each helper serves at a single nest for the duration of the breeding attempt but later may help at a different nest. These bee-eaters avoid incest and inbreeding by forming lifelong monogamous pairs with members of other clans. However, those who marry into another group do not totally separate from their natal clan; when not itself breeding, a bee-eater helps at a nest of its original clan instead of one belonging to the clan into which it has married, after which it may rejoin its mate. Bee-eaters prefer to aid their nearest kin.

At half of 194 nests monitored during three years, Emlen's team found from one to five helpers, with an average of 1.35 per nest. The proportion of the adult population engaged in breeding ranged from 15 percent in very dry years to 45 percent when conditions were more favorable. Many of the birds that refrained from breeding helped actively nesting pairs. With a life expectancy of about five years, a bee-eater may repeatedly change roles from breeder to nonbreeding auxiliary. Unassisted pairs produced an average of 0.53 young per nest. With one helper the yield was 1 fledgling; with two or more helpers, it rose to 1.34 fledglings per nest. In contrast to the situation among many other cooperative breeders, the amount of food delivered at a nest increased in direct proportion to the number of attendants; starvation of nestlings, a principal cause of their mortality, was reduced as the number of helpers rose. For about six weeks after their first flight, young bee-eaters are fed by their attendants.

Life in a busy colony of White-fronted Bee-eaters is not all peaceful cooperation. To avoid devoting ten to fourteen days to the laborious task of digging their own burrow, some pairs try to usurp the tunnels of members of other clans and may succeed after several days of fighting. When not closely attended by their mates, breeding females are chased and forced into coition by other males. About 17 percent of the nests are parasitized by females who surreptitiously lay in them. Sometimes this is done by a helper who will assist in caring for the nestling that hatches from her egg; more often the intruding female is a member of another clan who will not aid her host. When a foreign egg appears before the host female has started to lay, she may seize it in her bill and carry it out; if the egg appears while she herself is laying, she will incubate it and, with her helpers, raise the alien nestling.

In many ways, the White-fronted Bee-eaters' social system resembles that of communally breeding Chestnut-bellied Starlings (*Spreo pulcher*) and Gray-breasted and Southern San Blas jays (*Aphelocoma ultramarina* and *Cyanocorax sanblasianus sanblasianus*). The chief difference appears to be that the bee-eaters, constrained by the nature of their nest site to breed in crowded colonies, do not defend territories as the starlings and jays do, thereby avoiding some of the risks of colonial nesting. Even the continued association with her natal clan of a female who has married into another group is not without parallel in jays and other birds that do not nest in colonies.

About half a dozen other species of bee-eaters that inhabit savannas or other open country nest in colonies and breed cooperatively. In a colony of about twenty-five thousand Rosy Bee-eaters (*Merops malimbicus*) nesting in a sandbar in the middle of the Niger River, males were much more numerous than females, and up to six adults fed a brood. At a nest of White-throated Bee-eaters (*M. albicollis*) in the Sahel savanna at Lake Chad, six half-grown young were attended by seven grown birds. During an all-day watch, the presumed male parent brought food seventy-one times, the presumed female parent fifty-seven times, and the five helpers eighty-seven times. Although cooperative breeding is widespread among open-country bee-eaters, it appears not to be practiced by those that dwell in forests.

REFERENCES: Emlen 1978; Emlen and Demong 1984; Fry 1967, 1972.

15.
HORNBILLS
Bucerotidae

The forty-five species of hornbills range through Africa south of the Sahara down to the Cape and through much of India to southeastern Asia, Indonesia, the Philippines, and the Solomon Islands. These big birds, from 15 to 64 inches (38 to 163 centimeters) in total length, are best known for their great bills, which, unlike the equally long bills of the unrelated toucans of tropical America, are surmounted by a casque or excrescence that is often grotesquely large and curiously sculptured. Long-tailed, clad in black or brown boldly marked with white or cream, hornbills are seldom handsome. They are noisy birds, to whose loud, mostly unmelodious calls are added the strong swishing sounds of air rushing between the exposed bases of their wing feathers when they fly. Most are woodland birds that nest in holes in trees, but a few forage on the ground in savannas and other open country; they supplement their largely frugivorous diet with insects and small vertebrates. They lay from one to six white eggs, usually two to four, which the female incubates for four to six weeks. The slowly maturing young remain in the nest hole for another four to twelve weeks.

The Bushy-crested Hornbill (*Anorrhinus galeritus*) lives in the forests of Malaysia, Sumatra, and Borneo in groups of usually no more than ten birds, although parties twice as large have been recorded. It has a variety of whistling calls. Each group defends a territory of about 500 acres (200 hectares). When about to lay her two eggs, the female enters a cavity in a tree, made by decay or possibly by a large woodpecker. Then, probably by the combined efforts of the voluntary prisoner and her mate, the doorway is closed by a plaster of mud, regurgitated matter, and the birds' own excrements, leaving only a slit just wide enough to pass in food. Throughout the many weeks that the mother bird remains enclosed, she is fed by her mate with fruits, cockroaches, cicadas, and lizards. In all these details, Bushy-crested Hornbills differ little from the majority of hornbills, but in this cooperatively breeding species group members of both sexes and all ages help the breeding male provision the nest. Like most hornbills, the female molts and becomes flightless while safely enclosed, but she is able to repel predators that might try to break the hardened seal by thrusting her sharp bill through the narrow aperture. After ten or twelve weeks, the seal is broken and the female emerges with her two young, now well feathered and able to fly. During this long interval, the female and her nestlings have kept the cavity clean by ejecting their excreta directly through the aperture.

In Khao Yai National Park in Thailand, exposed to such hazards as tigers, aggressive elephants, king cobras, leeches, ticks, and stinging insects, Pilai Poonswad and her coworkers studied the nests of four species of hornbills. Only at a nest of the Brown Hornbill (*Ptilolaemus tickelli*) did they find helpers. Here food was brought to the enclosed female and her nestlings by six males of unknown age, five of whom were constant attendants, whereas the sixth arrived only occasionally. One of these males appeared to be the female's mate. The six attendants foraged and approached the nest together, waited until the last had delivered his food, then all flew off in a flock. After the fledglings emerged, one of the males was more closely attached to the female, while the others paid more attention to the young hornbills.

Largest of the hornbills, weighing up to about 9 pounds (4 kilograms), Southern Ground Hornbills (*Bucorvus leadbeateri*) inhabit savannas with scattered trees or bushes in Africa south of the equator. They have black plumage with white primaries, black legs and bills, a casque reduced to a simple ridge atop the upper mandible, and bare red skin on the face and inflatable throat pouch. The adult female is distinguished from the adult male by a patch of deep violet-blue on her chin. The only wholly carnivorous hornbills, they walk over the ground on the terminal joints of their toes, capturing snakes, lizards, frogs, snails, insects, and even mammals up to the size of hares. Much of their food is procured by digging or scratching it out of the earth with their strong bills. Despite their weight, they fly well and pursue invaders of their territory in high aerial chases. They roost together in trees with their heads tucked deeply between their shoulders and their bills pointing upward. On awaking at daybreak, the adult members of a group call all together, a booming *uuh u u h uuh-uuh* alternately low and high in pitch, the chorus so loud that on a still dawn it is audible nearly 3 miles (5 kilometers) away. Both immature and adult ground hornbills frequently play, often seizing each other's bills and wrestling in the manner of toucans, also tossing objects about with their bills, chasing one another on foot or in the air, repeatedly jumping on their companions' backs from a bank, tugging at their legs, or mischievously chasing other birds such as Helmeted Guineafowl (*Numida meleagris*). Like other cooperative

Southern Ground Hornbill

breeders, group members of both sexes, adult and immature, preen one another.

For five months, the Kemps watched fourteen groups of Southern Ground Hornbills in Kruger National Park in the Republic of South Africa. A group consisted of two to eight individuals, most often three to five, including adult males and females, subadults, and immature birds. These bands contained two or three adult males more often than two adult females. Females lived alone more often than males, but three stayed together with no male companion. Groups foraged over and defended areas of about 38 square miles (100 square kilometers) and each group contained a single dominant breeding pair. Although ground hornbills sometimes nest in crannies in cliffs or in burrows in earthen banks which they may dig themselves (apparently they are the only hornbills able to excavate their nest chambers), those in Kruger Park usually occupied large cavities in trees, from 13 to 23 feet (4 to 7 meters) above the ground. While a female sat in her chosen nest site, the males in her group came with food in their bills, then picked up grass and leaves for the nest lining and deposited the mixed load in the cavity. The female found and ate the food, then arranged the lining.

A female ground hornbill lays two eggs, which she incubates for about forty days. Unlike nearly all other hornbills, she is not sealed up in the nest, possibly because this protection, often effective high in forest trees, would avail little against the powerful predators of African savannas. The open nest permits her to escape if she is threatened. Unless it rains, she leaves her eggs about three times daily, for recesses of about thirty minutes, during which she preens and sometimes feeds herself. Most of her food is supplied by group members, chiefly by adult males, with immature birds contributing a little. The items, usually a number at a time if they are not large, are held conspicuously in the tip of the provider's great beak. One female was fed about eight times a day. When a meal was brought, the whole group, including members who had nothing to offer, approached the nest together. The same routine was followed after the chick hatched, with the difference that the mother herself also brought food to it after it was twenty-three days old.

The hatchling ground hornbill is covered with pink skin quite devoid of down. Its eyes are tightly closed and its upper mandible, tipped with an egg tooth, is shorter than the lower mandible, as in newly hatched kingfishers and woodpeckers. Within a week, the nestling becomes shiny purplish black. When fourteen days old its eyes have opened, and at three weeks pinfeathers are emerging through its skin. Unlike other hornbills, it does not keep its nest clean by defecating through the entrance. Developing slowly, it stays in the nest until about eighty-six days old. Thereafter follows a long period of immaturity, during which the young bird is distinguishable from adults by the color of its eyes and the bare skin of its throat. For about two years, it depends upon other group members for its food, and it does not become adult until it is six years old.

Because, even if both eggs hatch, one nestling always starves, a group of Southern Ground Hornbills rears at most one nestling per year. Moreover, groups often pass a year without breeding, and nests are destroyed by predators, with the result that the hornbills studied by the Kemps raised, on the average, only one fledgling per group every 6.3 years. In a stable population, such a low rate of recruitment must be compensated by longevity. The average life span of Southern Ground Hornbills past the vulnerable early months appears to be about twenty-eight years.

On several occasions, intruding individuals of the Casqued Hornbill (*Bycanistes subcylindricus*) offered food to a female of the same species enclosed in her nest hole in a tree. When the mate of the imprisoned female returned, he chased the trespassers away.

Nestlings of certain hornbills might be classified as helpers because they engage in activities, such as nest building, usually performed only by parent birds. After the emergence of their mother from a nest hole, two nestlings of the Red-and-White-billed Hornbill (*Lophoceros deckeni*), about three weeks old, replaced the plaster seal with materials they found inside the cavity, thereby enclosing themselves for the remainder of their residence in their natal chamber. Similarly, after the female Crowned Hornbill (*L. melanoleucos*) breaks out of the hole in which she has been sealed from the start of laying until her nestlings are half-grown, the young birds, working within their nest, plaster up the doorway again, leaving a gap just wide enough for their parents to pass food to them. Captive nestling Crowned Hornbills, placed in a box with a small opening, proceeded to reduce its size with mud supplied to them, particles of food, and their own excrement. They attended efficiently to the sanitation of their box, and one of them proffered food to its nest mates.

REFERENCES: Kemp 1979; Kemp and Kemp 1980; Kilham 1956; Moreau 1936; Moreau and Moreau 1940; Poonswad et al. 1983.

16.

PUFFBIRDS

Bucconidae

Thirty-two species of puffbirds, related to the woodpeckers and toucans, inhabit warm woodlands of the American continents from southern Mexico to northern Argentina. A heterogeneous assemblage, they are survivors of a lineage more widely distributed in past ages. A large head, short, stout body, and fluffy plumage give to many of these 5.5- to 11.5-inch (14 to 29 centimeters) birds a puffy aspect, like that of puffins. Never brilliant, they are clad in black, white, brown, rufous, and buff, often barred, streaked, or spotted. Their food consists almost wholly of insects and other invertebrates, which they pluck from bark or foliage or catch in the air on long sallies from perches. Their voices, thin and weak or loud and ringing, are rarely melodious. In chambers that they carve in termitaries, burrows in the ground, or, rarely, closed nests of other birds, unlined or lined with dead leaves or grass, puffbirds lay two or three, seldom four, unmarked white eggs that are incubated by both sexes.

Among the larger members of the family are the four species of nunbirds, named for the somber attire that all wear. The White-fronted Nunbird (*Monasa morphoeus*) ranges through the rain forests from eastern Honduras to Bolivia and southeastern Brazil. About 11 inches (28 centimeters) long, it has dark slate-colored upperparts and wing coverts, lighter gray underparts, and nearly black wing and tail plumes. The black heads of both sexes are adorned with short, stiff, outstanding white feathers on the forehead, lores, and chin. Projecting from this ruff is a bright orange-red bill, which seems as incongruous on a bird so somberly clad as lipstick on a nun.

In Costa Rica's Caribbean rain forests, where we studied them, White-fronted Nunbirds wander through the crowns of tall trees and across clearings in straggling parties of seldom more than half a dozen individuals. Often they associate loosely with other more or less social birds of the treetops, including Purple-throated Fruitcrows (*Querula purpurata*) and Scarlet-rumped Caciques (*Cacicus uropygialis*). Their flight is strong and direct but rarely long-continued. Like other puffbirds, they eschew fruits and subsist almost wholly upon insects and other small invertebrates, with rarely a little frog or lizard. They forage in the energy-saving manner that has given puffbirds an undeserved reputation for sloth and stupidity. Instead of flitting restlessly about, searching for food, as many birds do, they perch quietly on an exposed branch and watch intently. Suddenly they dart out, swiftly and directly, snatch a green insect from a green leaf, a brown roach from brown bark, or a spider from amid low herbage, perhaps 60 feet (18 meters) away, and carry it back to their perch, against which they may beat it before they swallow it. Adults feed one another, but allopreening was not seen.

Nunbirds have a great variety of notes, loud and full or soft and plaintive. The most impressive of their vocal performances is the chorus, given by three to ten birds who line up, close together but not in contact, on a slender horizontal branch or vine stretched between trees, from about 25 to 75 feet (8 to 23 meters) up in the midst of the forest or in an adjoining clearing. Tilting their heads upward, the dusky choristers call all together in loud, ringing, almost soprano voices, with such vehemence that their whole bodies shake. Often these bouts of simultaneous shouting, stirring if hardly melodious, continue for fifteen or twenty minutes, the volume of sound waxing and waning as more or fewer voices join in, but with only brief intermissions. By calling all together, the members of a breeding group appear to affirm their unity. Sometimes two groups join in a chorus; at the conclu-

White-fronted Nunbird

sion of a performance with eight participants, they separated into quartets that flew off in different directions through the forest.

White-fronted Nunbirds nest in burrows from 40 to 55 inches (100 to 140 centimeters) long in gently sloping or nearly level ground beneath tall forest trees. Apparently these tunnels are excavated by the birds themselves, although we did not see this. The expanded chamber at the inner end of the long tube is well lined with dead leaves. The burrow's mouth is framed by a thick collar of dead leaves, petioles, and coarse decaying twigs, which makes the opening less conspicuous amid the litter on the forest floor. Despite much searching during two breeding seasons, my son and I found only three nests, and we probably would have missed these if the adult birds had not directed our attention to them after the nestlings hatched.

Only in the shortest burrow could we count the nestlings. When first seen the three were a few days old, with closed eyes and pink skins devoid of down. At each burrow the young were fed by three or four nunbirds, all outwardly similar except for the tattered tails of some, which enabled us to recognize them individually. Unless it intended to brood, an attendant did not enter the burrow to feed even the smallest young. Flying up through the forest with a single item, usually an insect or spider, rarely a small frog or lizard, held conspicuously in the tip of its vivid bill, the nunbird never went directly to the burrow but alighted well above the ground in a nearby tree and repeated over and over a rippling or churring call while it scrutinized its surroundings. Assured that no enemy was in view, it descended to the tunnel's mouth and stood exposed while it delivered a rapid series of low, sharp notes. Alerted by this call, a nestling ran up to the entrance, took the food from the attendant's bill, then promptly scampered back to the chamber at the inner end. Even the youngest nestlings toddled up the gently inclined entrance tube, bobbing their sightless heads and waving their rudimentary wings, received their meal, then returned to the warmth of a brooding parent. The pads of thickened skin on the nestlings' heels, which seemed superfluous in the chamber well lined with leaves, protected these joints from abrasion on the round trip of 8 or 9 feet (2.4 to 2.7 meters) over the earthen floor of the longer burrows.

Unlike many forest birds, the nunbirds were surprisingly tolerant of the human presence and could be watched at their nests without concealment from a distance of a few yards. They were never seen to remove waste from a burrow, where maggots soon battened in the bed of dead leaves. Probably these wriggling larvae helped break down waste and make the burrows less odorous. One morning I returned to the shortest burrow to be greeted by a gaping hole with a pile of freshly dug earth in front. Some powerful mammal, probably a Tayra (*Eira barbara*), had taken the three nearly feathered nestlings. For at least twenty-four hours after their loss, the four attendants continued to bring food, stand on the mound of soil that the pillager had made, and call the missing nestlings to come and get it.

When they leave the burrow at the age of about thirty days, young nunbirds rise high into the trees. Soon they take food in a spectacular fashion, flying up from a distance to snatch it from an attendant's bill while on the wing, thereby gaining skill in the nunbird's habitual manner of foraging. Neither they nor the adults return to sleep in the burrow. Apparently, the young remain with their parents into the following year, when they help raise another brood.

REFERENCE: Skutch 1972.

17.

BARBETS

Capitonidae

Most of the seventy-six species of barbets live in the eastern hemisphere, from Africa south of the Sahara to India, Malaya, the Philippines, Sumatra, and Borneo. Africa is especially rich in barbets, but only thirteen species inhabit the New World, from Costa Rica to Brazil and Bolivia. Barbets are short-necked, stocky birds from 3.5 to 12 inches (9 to 30 centimeters) long. Like their relatives the woodpeckers, toucans, and puffbirds, they are zygodactyl, with two toes directed forward and two rearward. At the base of their short, stout bills are the bristles that suggested the family name. Their plumage is often brilliant, even gaudy, with red, yellow, green, and blue, in addition to black and white, adorning the same small bird; a few are clad in sober shades. Many consume vast quantities of figs and other soft fruits, plus a few insects, in the crowns of forest trees; but on the savannas of Africa, some hunt over the ground for termites and other small invertebrates.

Barbets have strong, not always melodious voices, which they use freely in duets by pairs or choruses by groups—the metallic timbre of the notes of some has earned them such sobriquets as coppersmith, blacksmith, and tinkerbird. Most barbets nest in unlined, woodpeckerlike holes that both sexes carve in dead trunks and branches, but a few dig burrows in banks or even vertical shafts in level ground, with the brood chamber well below the surface. Their two to four plain white eggs are incubated by both sexes and hatch in twelve or thirteen days. The nestlings, downless at birth, are fed by both parents, at first with insects but soon with much fruit.

Nest helpers are frequent among barbets but, probably because their nests in decaying trees of uncertain stability are difficult or hazardous to reach, we know too little about them. In South Africa, Carl Vernon found that a group of Black-collared Barbets (*Lybius torquatus*) defended a territory throughout the year, and in the nonbreeding season all members slept together in the same nest hole. To breed, a group of eight birds split into subgroups of three, three, and two, each with its own nest. The whole group continued to defend its territory, and the trios or pair visited each others' nest trees. At a nest of the Pied Barbet (*L. albicauda*) in Kenya, V. G. L. Van Someren watched four birds, all similar in appearance, bring food to an undetermined number of nestlings. Once he found six grown birds sleeping together in a tree hole. In the Frankfurt Zoo, nestlings of the Double-toothed Barbet (*L. bidentatus*) were fed by a second female

Red-and-Yellow Barbet

and juveniles of the preceding brood. After the young fledged, the whole group continued to feed them, and they were led back to sleep in the nest.

At a nest hole of the White-eared Barbet (*Stactolaema leucotis*) in Tanganyika (now Tanzania), the Moreaus watched four adults bring thirty large berries to four nestlings, in ninety minutes, and remove droppings nine times. Since other nests of this species contained only two or three eggs, they surmised that two females had laid in this one. At another nest of this barbet, an extra adult occasionally brought food to the young birds and slept in the nest hole with them and their parents. Many years later, Lester Short and Jennifer Horne watched three nests of White-eared Barbets situated inaccessibly high in dead stubs in Kenya. At one of these nests five birds shared incubation, and at another six participated in this office. To the third nest six adults brought food ten to eighteen times per hour (average 15.4 times per hour) for at least two nestlings, and all six slept in the nest hole with them. In South Africa, eleven birds of this species roosted together in a cavity.

In arid East Africa, colorful Red-and-Yellow Barbets (*Trachyphonus erythrocephalus*) forage over the ground and nest and sleep in burrows that they dig in

Double-toothed Barbet

Prong-billed Barbet

the walls of ravines or in tall termite mounds. Three adults reared the young in one nest and four adults in another.

The only New World member of the family that has been somewhat thoroughly studied is the thick-billed, plainly attired Prong-billed Barbet (*Semnornis frantzii*), which is confined to the mountain forests of Costa Rica and western Panama. In the breeding season, pairs are strongly territorial; at nests that I watched during excavation, incubation, and rearing the young, only the two parents were present. Both slept in the nest hole while it held eggs and nestlings, and after the young fledged they were led back to lodge in it with their parents. After the breeding season, families joined to pass the night in cavities high in trees. Sixteen grown birds entered a hole so small that they must have slept in closest contact in several tiers, thereby revealing a habit widespread among cooperative breeders, many of which are "contact birds," and suggesting that their ancestors may have been more sociable throughout the year than the birds whose nesting I watched.

REFERENCES: L. G. Grimes 1976; Moreau and Moreau 1937; Reynolds 1974; Schifter 1972; Short and Horne 1980, 1984; Skutch 1983a; Van Someren 1956; C. J. Vernon in L. G. Grimes 1976.

18.
TOUCANS
Ramphastidae

Thirty-seven species of great-billed toucans, related to the woodpeckers and barbets, inhabit the wooded regions of the tropical American mainland from southern Mexico to Paraguay and northern Argentina. One of the plainest members of the family is the middle-sized Collared Araçari (*Pteroglossus torquatus*), which ranges from southern Mexico to northern Colombia and northwestern Venezuela. About 16 inches (41 centimeters) long, it is largely blackish, with a bright red rump and a yellow breast separated from the yellow posterior underparts by a broad black band. Its yellow eyes are set amid bare red skin; its long bill, less vividly tinted than many toucans' bills, has a black lower mandible and a dull whitish upper mandible. The sexes are difficult to distinguish. Like other toucans, araçaris are largely frugivorous but vary their diet with insects and occasional eggs or nestlings of other birds.

As I wandered through the darkening forest of Barro Colorado Island in central Panama on a February evening, long ago, a sharp *penk* drew my attention to a flock of Collared Araçaris in the treetops high above me. They were trying to enter a cavity in a thick horizontal limb, about 100 feet (30 meters) up in a gigantic tree. It was not easy for them to pass through the narrow doorway that faced directly downward. As I watched their dark, colorless figures, silhouetted

Collared Araçari

against the last glow of daylight, they fluttered below the hollow bough and frequently turned back, to try again and again until they secured a foothold on the edge of the orifice and each squeezed laboriously in. At the following dawn, six araçaris struggled outward, one by one, before all flew off through the treetops.

After the six araçaris had slept in the high dormitory for several weeks, their number was gradually reduced, until by the end of March only one remained. Now the hole was occupied by day as well as by night. Whenever I stood in view and loudly clapped my hands, a big pied bill was thrust through the narrow aperture; then a slender body slowly followed, to fly beyond view. Despite its great height, the incubating bird did not feel safe in my presence. This continued for about three weeks, during which the other members of the group lodged elsewhere, leaving the incubating parent alone with the eggs, which might have been broken in a crowded dormitory.

By the second week of April, five araçaris again slept in the high hole, the sixth having inexplicably vanished. Now they were feeding nestlings, but so shy and distrustful were they that I had to watch from a blind, as I have seldom found it necessary to do at such high nests of other kinds of birds. Since these araçaris all looked alike, I could tell how many fed the young only when all five did so in rapid succession; but I could not make sure that more than two brooded. In toucans that nest without helpers, the male and female share rather equally incubating, brooding, and feeding nestlings.

Instead of flying directly to the nest in a confident manner, an attendant arriving with food would perch in a neighboring tree, turning its head and great bill from side to side to scrutinize its surroundings before it approached the doorway. This pause often enabled me to learn what it brought. For some days after the nestlings hatched, I noticed only insects, grasped in the tip of the bill, wings and legs sometimes protruding. Even when the nestlings were a month old, they still received many winged insects, but fruits were becoming more prominent in their diet. Now that larger quantities were needed for the growing araçaris, most items were carried in the attendants' mouths or throats, and I rarely saw them.

For over a month, the adults wriggled into the cavity each time they came with food. At least thirty-five days after the nestlings hatched, I first saw them take some of their meals through the doorway, while the feeder clung beneath it. In the following days, an increasing proportion was passed to them in this fashion, easing the attendants' task. Now the adults entered by day chiefly to remove big billfuls of waste, including large regurgitated seeds of the "wild nutmeg," a species of *Virola*, from which the bright red, branching aril had been digested by the young. All five adults continued to sleep with the nestlings in the capacious cavity.

Thirty-seven days after I first saw an adult take food to the nest, a nestling pushed its head outside to receive its meal. Its bill looked almost as big as that of an adult. When the young were at least forty-two days of age, one or another spent much time looking through the doorway, and I first heard them call *pitit*, like the adults but more weakly. The attendants now became even more excitable and distrustful than formerly—with good reason, as I was soon to learn.

On May 24, a fledgling at least forty-three days old emerged from the high nest. I did not witness its departure, but in the evening of that day I watched the adults help it return to the hole for the night. Difficult for the old birds, it was doubly hard for the inexperienced fledgling to enter. While an attendant hung, back downward, beneath the doorway, the youngster clung momentarily to its back. Meanwhile, the other adults clustered around, crying *pitit, pitit* in high-pitched voices and displaying much excitement. The foliage was fast falling from the nest tree, leaving the araçaris exposed to the open sky. Without warning, a big White Hawk (*Leucopternis albicollis*) swooped down and seized the fledgling in its talons. The victim's piteous cries, like the wailing of a disconsolate child, grew fainter in the distance as the raptor bore it across a deep ravine, followed by all the adults, who rallied to its defense, much as toucans of a different species gathered around Walter Bates when he picked up a member of their party that he had shot in Amazonia. Later, when the sky had grown dim, all five adult birds cautiously returned to the nest tree and pushed into the cavity as rapidly as the narrow orifice would permit.

In the next three days, at least two more fledglings emerged from the high hole, but they were not led back to sleep in it. Doubtless, the adult araçaris found a safer, more easily entered dormitory for them. Nevertheless, until at least early June, four adults continued to lodge in the cavity where the brood had been raised. Since toucans seem to rear only a single brood in a season, I concluded that the three helpers at this nest were yearlings if not older.

Nests of many rain forest birds are exceedingly difficult to find. In the half-century since I watched this family in the Panamanian forest, I have seen only one other nest of the Collared Araçari, in drier woodland, too distant to be studied. A few more nests have been reported by others, but none appears to have been carefully watched. Nevertheless, since nesting birds follow patterns widespread in their species, I doubt that this cooperative family of Collared Araçaris was unique. In the toucan family as a whole, however, cooperative breeding may be exceptional, as studies of a number of other species in three genera have failed to find helpers. But the thirty species of toucans that await careful study should yield some surprises to diligent observers.

References: Skutch 1958, 1983a.

19.

WOODPECKERS

Picidae

One or more of the 204 species of woodpeckers is found almost everywhere that trees grow, except Madagascar, New Guinea, Australia, New Zealand, and small islands remote from continents. A few inhabit treeless regions, such as the Argentine pampas and the high Andes above timberline, where they walk over the ground. In size they range from the 3.5-inch (9-centimeter) piculets, of which 27 species inhabit tropical America, to the Imperial Woodpecker (*Campephilus imperialis*) of Mexico, 22 inches (56 centimeters) long. In coloration they are extremely varied and often quite handsome in black, white, red, yellow, and golden-green, frequently barred or spotted. The sexes are usually alike except on the head, where the male often has more red than the female, making it easy to distinguish them. Although highly specialized for climbing over and digging into tree trunks in search of insects, many woodpeckers are versatile foragers that catch insects in flight and gather fruits or nuts. Nearly all carve holes in trees in which they lodge at all seasons, singly or in family groups. On the unlined floors of these cavities the females lay from two to eight glossy, porcelain white eggs, which the sexes incubate alternately by day and the male, usually the more domestic member of the pair, warms through the night. After an incubation period of eleven to eighteen days, the

Red-cockaded Woodpecker

downless nestlings hatch. Fed and brooded by both parents, they remain in the cavity for twenty-one to thirty-five days.

The Red-cockaded (*Picoides borealis*) is in many ways an unusual woodpecker. Although the sexes of most members of the family are readily distinguishable by their head markings, the spot of red on each side of the male's nape, for which the species is named, is so tiny that it often escapes detection in the field. Aside from the male's inconspicuous decoration, the sexes of this small woodpecker wear the same colors: black on the crown, tail, and white-barred back and wings, white on the cheeks, outer tail feathers, and underparts, with black streaks on the sides and flanks. Juvenile males, browner than adults, wear a small, round patch of red on the center of the crown that is lacking on juvenile females and adults of both sexes.

Pine trees are as indispensable for Red-cockaded Woodpeckers as oaks are for Acorn Woodpeckers (*Melanerpes formicivorus*). They inhabit open, mature pine woodlands in the southeastern quarter of the United States, from eastern Oklahoma and southern Maryland to Florida and eastern Texas. Such open stands of pines, sometimes mixed with deciduous trees, are perpetuated by recurrent ground fires. Modern forestry management, which prevents fires and clear-cuts the pine woods, has been altering or destroying the Red-cockade's habitat so rapidly that it may be an endangered species. Controlled burning as

Acorn Woodpecker

a method to maintain special habitats, as well as the preservation of unmanaged pine forests, offers hope for survival.

Red-cockaded Woodpeckers need mature pine woods because only there can they find enough older trees, infected with the red heart rot caused by the fungus *Fomes pini*, in which they can carve holes for sleeping and nesting. Penetrating the bark and living sapwood, the woodpeckers excavate downward into the softened heartwood free of resin to make a hole 8 to 11 inches (20 to 28 centimeters) deep. Around the circular doorway of each hole, they peck many small pits into the bark and underlying sapwood, causing resin to flow abundantly for several feet above and around the orifice. To keep the resin fresh and sticky, the woodpeckers continue to perforate the bark, usually before they enter the cavity for the night. As far as is known, no other woodpecker has this peculiar habit, the purpose of which puzzled ornithologists until Jerome A. Jackson demonstrated experimentally that it holds aloof the abundant Gray Rat Snakes (*Elaphe obsoleta spiloides*) which prey heavily on the eggs and nestlings of birds. Snakes that became extensively smeared with pine resin sickened and sometimes died.

In these protected dormitory and nest cavities, pairs or clans of Red-cockaded Woodpeckers sleep singly throughout the year. By day they wander over home ranges of about 42 to 173 acres (17 to 70 hectares), gleaning insects and spiders from the bark of pines and sometimes other trees. As in certain other woodpeckers, the sexes forage on different parts of the trees, males usually high on trunks and branches, females lower on the trunks—a separation that reduces competition for food and promotes more efficient use of resources. Berries supplement the woodpeckers' diet. In the evening, each retires into its own roost hole; unlike certain other woodpeckers, adult Red-cockades appear never to share a dormitory. Those who lack proper holes protected by fresh resin may roost clinging high in a pine tree, occupy an old hole around which the resin has dried, or fly to a neighboring territory to lodge in a hole in a living pine that may be over a mile away from their own nest tree. Since Red-cockaded Woodpeckers need from ten months to several years to complete a cavity in a living pine, they may continue for well over a year to sleep in a foreign territory. To avoid being chased by the resident clan, the intruding woodpeckers usually enter their borrowed holes after the residents retire in the evening and leave before they become active in the morning.

In winter, groups or clans of Red-cockaded Woodpeckers consist of two to eight birds, including young of the latest brood. In April, when nesting begins, the larger groups are reduced to a single breeding pair plus one to three helpers, most often males who are usually yearling or older offspring of the mated pair. Some assist the same pair for at least two breeding seasons. No special cavity is carved for the brood; the female lays three or four eggs on the unlined floor of a dormitory, nearly always her mate's rather than her own. During the surprisingly short incubation period of ten days plus a few hours, the auxiliaries help warm the eggs. The male parent usually spends somewhat more of the daytime on the eggs than does the female, and he incubates them through the night. Together, two helpers may contribute more hours to diurnal incubation than either of the parents. The male parent tends to feed the nestlings somewhat more frequently than their mother, but two or three helpers together bring food more often than either parent. Helpers as well as parents brood the nestlings, each, after delivering food, usually remaining in the cavity until another arrives with a meal. The father is chiefly responsible for nest sanitation, carrying out nearly all the fecal sacs, with a little help from his mate and very little from their assistants. All join in defending the nest.

When about four weeks old, the fledglings leave the nest cavity and thereafter are fed and guarded by parents and helpers. They continue to receive food from the adults for five or six months—far longer than most woodpeckers. In another aspect of parental care, Red-cockades are less solicitous than certain less social woodpeckers. Instead of installing newly fledged young in a snug cavity as evening approaches, the adults leave them to shift for themselves. For several weeks, they roost high on the trunks of pine trees. Or they may find and sleep in holes in dead trees until they can carve or otherwise acquire a cavity in a living pine.

Eleven nests without helpers fledged an average of 1.7 young, whereas thirteen nests with helpers produced 2.1 fledglings per nest. Unaided parents raised slightly more fledglings for each adult attendant than trios did.

The Acorn Woodpecker is no less unique than the Red-cockaded, but in different ways. As closely dependent upon oak trees as the Red-cockaded is upon pines, it lives among oaks from Oregon through California to southern Baja California and from Arizona to western Texas, thence southward through the highlands to western Panama, with an isolated race in the Andes of northern Colombia. The Acorn Woodpecker wears a harlequin pattern of white and glossy greenish black, with a red cap and bright yellow eyes. Shouting *rack-up rack-up* while it bows deeply up and down atop some lofty stub or *r-r-r-rack-up, r-r-r-rack-up* with a long, deep roll, this sociable woodpecker impresses one as an exceptionally jolly bird, and certainly it is one of the most amusing to watch.

In autumn when acorns are ripening, these woodpeckers are extremely active, plucking the nuts from the trees and driving each into a hole of appropriate size that they have carved in a dead trunk or the thick bark of a living tree, which may be honeycombed with thousands of these little depressions, each with its closely fitting acorn. Although other woodpeckers and birds of different families store food, no other is known to have such a neat way of doing it. Well known in coastal California, these granaries have been found in scattered localities from Mexico to Panama, but they are not prepared by all populations. In the highlands of Guatemala and Costa Rica, I have watched Acorn Woodpeckers simply tuck whole or fragmented acorns into crevices in wood or bark or amid massed epiphytes. Their stores of mast are especially important to Acorn Woodpeckers in the north where winters are cold, and successful breeding may depend upon having enough to last until the following spring. However, acorns are by no means the woodpeckers' only food. They climb over trunks and limbs of trees, extracting small creatures from bark and wood; they are skillful aerial flycatchers; and, like sapsuckers (*Sphyrapicus*), they drink sap from little pits that they drill into the bark of living trees.

Unless forced to wander by the exhaustion of their food supplies, Acorn Woodpeckers are highly sedentary. In a grove of oaks about 2 miles (3 kilometers) from my Costa Rican home and 500 feet (150 meters) higher, a colony of these birds flourished for years, yet I have never seen a single one on our farm, where no oaks grow in the forest. Acorn Woodpeckers live at all seasons in groups of two to fifteen individuals, including one to four sibling males, one or two sibling females, and their offspring from one or more preceding years. Each group defends a territory which contains oaks and sap trees for food, granaries for storage (at least in northern and some tropical populations), nest holes, and dormitory holes. Some of the latter are occupied year after year and may later serve as nest holes. Up to twelve individuals, the entire membership of a group, have slept in a single hole, but frequently a large group occupies several cavities, a few in one and a few or perhaps a single woodpecker in another. Alarmed by the approach of a hawk, the woodpeckers sometimes take refuge in their holes. In rainy weather, they spend much time in them, but more often they cling beneath branches for shelter. Unlike many cooperative breeders, Acorn Woodpeckers appear not to preen one another or to rest in contact by day; but when they crowd into a dormitory, they can hardly avoid sleeping close together. Although dominance relations exist, they are rarely revealed by overt behavior.

Each group or territory has a single active breeding nest at one time. Most nests of Acorn Woodpeckers contain three to five eggs, with four being the usual number, but as many as thirteen have been found. The great variation in clutch size was puzzling until it was ascertained that two eggs are sometimes laid in the same nest on the same day, which one female is unlikely to do. About one-third of the nests have more than five eggs, apparently the product of two females, and about half of all breeding females nest communally. When two share a nest, one sometimes throws out some of the other's eggs.

Whether the nest is communal or contains eggs of a single female, in groups consisting of more than one male and one female it is usually attended by helpers, including yearlings and two-year-olds. When several birds incubate, changeovers can be surprisingly frequent. At a nest belonging to a group of five, inaccessibly high in a charred trunk on a Costa Rican mountainside, the attendants replaced each other every few minutes; 108 sessions by all cooperating woodpeckers averaged only 5.1 minutes, with a minimum of less than 1 and a maximum of 17 minutes. A single male incubated the eggs by night.

Although the auxiliaries at nests of Red-cockaded Woodpeckers are nearly always males, both sexes are well represented among those at Acorn Woodpeckers' nests. A larger proportion of female nonbreeding helpers than of male nonbreeding helpers incubate, brood, and feed nestlings, and the former do so more actively. Although yearlings engage in all these occupations, they do less than older helpers. The young birds also participate in defending territories, storing acorns, and carving holes. Some yearling and older members of a group may fail to assist at a nest; and when, as rarely occurs, a second brood is reared, juveniles of the earlier brood do not help. In this, Acorn Woodpeckers differ from many other cooperative breeders.

I have found no statement as to whether, like Golden-naped Woodpeckers (*Melanerpes chrysauchen*) and piculets, Acorn Woodpeckers lead their newly emerged fledglings to sleep in holes or whether they must find dormitories for themselves. For about two weeks after they begin to fly, fledgling Acorn Woodpeckers are fed by most of the adults and by yearlings, who give them less than the older birds do. After this interval, the juveniles support themselves so adequately with sap from the wells, supplemented by insects that they catch in the air, that they need little from their attendants. By late summer, when green acorns become the woodpeckers' principal food, the juveniles are so inept at gathering, carrying, and opening them that they are again largely dependent upon their elders for nourishment. By autumn, when the older birds are busily storing acorns, most ju-

Golden-naped Woodpecker

veniles can feed themselves. However, for obscure reasons, some of the young birds are denied access to the granaries by their elders, who during the winter months in California supply them with acorns, sometimes continuing until the recipients are ten months old—an exceptionally prolonged period of juvenile dependence.

The Acorn Woodpeckers' social system, which permits both males and females to breed in the group that reared them, might lead to deleterious inbreeding but for two compensating factors. In the first place, although they may become reproductively mature when one year old, these woodpeckers do not breed in their natal group until all possible parents of the opposite sex have disappeared and been replaced by outsiders—a limitation that implies the ability to recognize individuals and to remember them well. Second, these replacements are subordinate members of other groups who emigrate singly or with companions of the same sex, either two females or two to four males together, never a male and a female together, to seek an opening elsewhere.

The wanderers are especially attracted to groups from which the dominant individual of either sex has disappeared. Competition to fill such vacancies is often intense; the attempt of aliens to establish themselves in a group frequently attracts other individuals with similar inclinations, leading to power struggles which are most often witnessed as nesting begins in April. As the intruders strive with each other for the coveted position and the resident woodpeckers try to expel them, excitement rises to the highest pitch. Rarely as many as forty individuals participate in such a melee, calling loudly, chasing each other, grappling and falling to the ground locked together. In the same territory, the struggles may be renewed at intervals over a period of two or three weeks. Usually they end with the incorporation of a new member, more often female than male, into the group. Although in other cooperative breeders outsiders who try to enter a group are resisted, fighting on the scale that erupts among Acorn Woodpeckers has not been reported except possibly in Noisy Miners (*Manorina melanocephala*).

The Acorn Woodpeckers' cooperative breeding system does not promote maximum reproduction. Although solitary pairs, more numerous than larger groups, produced the most fledglings per adult attendant, most individuals were active at cooperative nests. At the Hastings Natural History Reservation in Monterey County, California, groups of seven or eight birds were the most productive, but those with five or six members were more numerous. Studies in different localities in different years revealed that Acorn Woodpeckers raised to the age of nest leaving from 0.2 to 0.3 young per adult per year, which is one fledgling for every five or every three adults in the population. This low reproductive rate has several causes. Nearly half the groups may allow a year to pass without breeding, and many nests fail. Moreover, about half the fledged young disappear before they are a year old. Only a fairly long-lived bird could continue to flourish with such low recruitment. Adult males, who predominate in the population, have an estimated life span of thirteen years, females one of about ten years.

One theory of the origin of cooperative breeding (which we shall consider in more detail in chapter 53) attributes it to habitat saturation: a population of birds so completely fills the available habitat that emigrating progeny are unlikely to find an area where they can survive, much less breed; accordingly, they stay at home and help their parents defend the family territory and rear later broods. In this connection it is interesting to notice that, in contrast to the sedentary, cooperatively breeding Acorn Woodpeckers of other regions, those that nest in southeastern Arizona are migratory, passing the winter in Mexico. Like most migratory birds, they do not travel in family groups. When they return individually in the spring, they frequently take new territories and new partners, to nest in simple pairs without helpers.

Intermediate between these migratory Acorn Woodpeckers and those of coastal California that have chiefly claimed our attention is the population in Water Canyon in the Magdalena Mountains of central New Mexico, studied for many years by Peter Stacey. Here the birds lived in the valley bottom, where most of the oaks and other large trees that supplied food, storage, and nesting sites grew. Their territories stretched in linear sequence along the stream and were open-ended on the pine-clad slopes of the canyon. Between such narrow valleys where they could thrive were ridges and high desert grasslands unsuited for them. In the canyon, the woodpeckers stored acorns and seeds of the Pinyon Pine (*Pinus edulis*) in typical granaries, chiefly in the dead parts of cottonwood and pine trees. When, as happens at intervals in the southwest of the United States, the nut crop fails, the woodpeckers drill no storage holes and may be forced to wander in search of food.

In this less favorable habitat, some territories were without woodpeckers at the start of the breeding season, although at the Hastings Reservation in coastal California such vacant territories were not found. Likewise, groups in Water Canyon were smaller than those in California, with only two to five members (an average of 3) instead of two to twelve (an average of 4.8). Groups with only two adults were more frequent than in California. The groups in Water Canyon were much less stable: juveniles more frequently left their natal group to join another group or to vanish from the study area; adults were less likely to nest in the same territory in successive years. Associated with so much shifting about, and possibly its cause, was the frequent presence of groups that lacked a breeding male or female and were, accordingly, inviting to newcomers. If the immigrants arrived before breeding started, they helped attend the nest and young; if they appeared afterward, they failed to assist, possibly because they could not have procreated any of the brood, although such aloofness is rare in cooperatively breeding birds.

In the less saturated habitat of Water Canyon, the woodpeckers raised more young than at the crowded Hastings Reservation, fledging an average of 2.5 instead of 1.2 offspring per group per year and 0.8 instead of 0.3 young per adult. Moreover, second broods, although nowhere common, were five times more frequent at Water Canyon than in California; and nearly twice as many groups fledged at least one young. However, successful nests yielded the same number of fledglings, an average of 2.7, in the two localities. The contributions to this result by groups of different sizes varied greatly. Simple pairs produced an average of only 1.26 young per year, while groups of three to five fledged 3.16 young per year—2.5 times as many. Larger groups were much less likely to lose their nests; in three years, only one of twenty-five groups of three, four, or five woodpeckers failed to rear at least one young bird, but only eight of fifteen simple pairs nested successfully. Only groups of three or four occasionally reared second broods; simple pairs, never. However, groups of four or five Acorn Woodpeckers did not rear significantly more young than groups of only three, which again indicates that these woodpeckers fail to reproduce as rapidly as they might do if they nested in smaller groups.

Golden-naped Woodpeckers inhabit the shrinking rain forests of the southern half of Costa Rica's Pacific slope and adjacent Panama. They are black and white, with a red abdominal patch. The male's red crown separates his yellow forehead from his yellow nape; the female differs from the male only in her black crown. High in massive dead trunks of forest trees or sometimes in the lower trees that shade a nearby coffee plantation, these woodpeckers carve deep holes in which they reside throughout the year. During the twelve days that they incubate their three or four white eggs and the five weeks that the young remain in the nest, the male and female nearly always pass the night together in the nest cavity—an arrangement unusual among woodpeckers. After the young fly, the parents lead them back to sleep in the hole. If all goes well, the family of two parents with two or three self-supporting young continues to lodge together until the following breeding season, when the parents move into a newly carved hole and their offspring disperse.

In the Valley of El General, where I have studied Golden-napes for many years, they only exceptionally raise two broods in a single season. When this occurs, juveniles of the first brood lodge with their parents while the latter incubate the second set of eggs and attend the nestlings. At one nest, two or three females of the earlier brood, when nearly four months old, brought food to their younger siblings, sometimes in the face of mild opposition from their mother; but they delivered it inexpertly and their contributions were negligible. After the second brood fledged, the family, consisting of the two parents, three females of the first brood hatched late in March, and one male and one female of the second brood hatched late in June, slept in the nest cavity. When the branch with this hole fell, the younger members of this family of seven helped the mated pair carve a new one. At the age of fifty-seven days, the male of the second brood lent his bill to the work. He continued to receive food from his parents until at least ninety-four days old. Although Golden-napes are among the more social woodpeckers, and juveniles rather ineffectually help feed younger siblings and more expertly help carve holes, the departure of the yearlings just before their mother lays in the following season prevents their becoming true cooperative breeders.

Brief observations of the closely related Yellow-tufted Woodpecker (*Melanerpes cruentatus*) of northern South America suggest that it has advanced further in this direction. One female successively entered three occupied nest holes close together in eastern Peru, and three adults entered one of these cavities. Pairs of handsome Hispaniolan Woodpeckers (*M. striatus*) may nest alone, or up to two dozen pairs may nest colonially, often in trunks of royal palms that also support the huge, many-chambered nests of Palm Chats (*Dulus dominicus*). A male and two females, one of them possibly a helper, entered the same hole. Careful studies may reveal that these and other tropical woodpeckers breed cooperatively.

Sapsuckers nest chiefly where severe winters discourage the permanent residence of united families that is the usual prerequisite of cooperative breeding, but they provide examples of altruistic helpers. When a male Yellow-bellied Sapsucker (*Sphyrapicus varius*) was killed by a Raccoon (*Procyon lotor*) while brooding nestlings, his mate continued to feed them. Soon she attracted a new male who, after a day of courtship, fed her nestlings and brooded them by night. This cost the new male his life, when the Raccoon returned in the dark and killed both him and the nestlings. Unmated male Williamson's Sapsuckers (*Sphyrapicus thyroideus*) frequently looked into nests of neighboring pairs whose nearly fledged young were calling loudly for food. Despite being chased by intolerant parents, some of these intruders fed the nestlings. A female whose mate vanished the day before her young fledged was joined by a bachelor who fed her brood both in the nest and after they emerged—behavior which contrasts sharply with that of certain aggressive male mammals, including lions, langurs, and some human tribes, which, when they become attached to a female with sucklings, destroy them so that she may sooner become pregnant with their offspring. The birds' habit of feeding offspring with food from their own mouths, sometimes when the feeder is obviously hungry, is more conducive to altruism and abnegation than the mammalian system of suckling the young.

Approached by a persistently begging newly fledged Tufted Titmouse (*Parus bicolor*), a Red-bellied Woodpecker (*Melanerpes carolinus*) carrying food for its own fledgling delivered it to the young titmouse. Apparently the woodpecker did not repeat the feeding. Marilyn Shy found three other cases of interspecific feeding by woodpeckers. A male Hairy Woodpecker (*Picoides villosus*) gave food to the calling fledglings of a Downy Woodpecker (*P. pubescens*). A mateless female Three-toed Woodpecker (*P. tridactylus*) fed the calling nestlings of a Black-backed Woodpecker (*P. arcticus*). A female Northern Flicker (*Colaptes auratus*) brought food to nestling European Starlings (*Sturnus vulgaris*).

References: Crockett and Hansley 1978; Curry 1969; Hooper and Lennartz 1983; Jackson 1974; Joste et al. 1982; Kilham 1977; Koenig 1981a, 1981b; Koenig et al. 1983; Koenig and Pitelka 1979; Lay and Russell 1970; Lennartz and Harlow 1979; Ligon 1970; Sherrill and Case 1980; Short 1970, 1974; Shy 1982; Skutch 1969b, 1980; Stacey 1979a, 1979b, 1981; Troetschler 1976.

20.

OVENBIRDS

Furnariidae

The 214 species of ovenbirds spread widely over the tropical and temperate zones of South America but to the north are found only in Central America and tropical Mexico. They are small or middle-sized passerine birds of the most varied habits, dwelling in heavy rain forests, dense thickets, arid woodlands, grasslands, and marshes. Almost wholly insectivorous, they seek their food by searching among the foliage of trees, creeping over trunks and branches, hopping through thickets, or walking over the ground. Lacking bright spectral colors, they are attractively clad in brown, rufous, chestnut, buff, gray, black, and white, often barred, streaked, or spotted. The family is named for the ovenbirds of the genus *Furnarius*, of which the best known is the Rufous Hornero (*F. rufus*) of Argentina and southern Brazil, which builds on a stout branch, fence post, or other firm support a solid structure of hardened mud or clay that resembles the domed baking ovens widespread in Latin America. However, no other family of birds constructs a greater or more fascinating variety of nests, which, if not situated in a cavity in a tree, burrow in a bank, or crevice among rocks, are covered—never, except in one species, simple open cups. Advanced cooperative breeding is unknown in this great family, most of whose members remain to be carefully studied, but in several species juveniles help build.

Rufous Hornero

In Argentina, Rufous Horneros live throughout the year in monogamous pairs on defended territories of about 0.6 to 2.5 acres (0.25 to 1 hectare), where they walk over the ground searching for insects, earthworms, and other invertebrates. Both sexes build the oven of clay, which does not open directly to the outside but is entered through a curved vestibule or antechamber. After building for two or three months, the horneros line the structure with dry grasses and a few feathers, upon which the female lays, at two-day intervals, three or four of the white, unmarked eggs usual in the ovenbird family. Both sexes incubate for sixteen or seventeen days, and both feed the downless nestlings. Their massive nest protects them well, for in a study by Rosendo Fraga 83 of 115 eggs, or 72.2 percent, produced fledglings—an exceptionally good record. After leaving the nest when twenty-four to twenty-six days old, young horneros soon become self-supporting, but they remain on the parental territory for four to nine months. At two nests, Fraga watched juveniles helping their parents build. At one of these nests, two young birds appeared to be doing a substantial share of the work. If, when they resume breeding, most parent horneros did not become intolerant of the presence of juveniles near their nests, they might more often receive help in building and perhaps even in feeding subsequent broods.

The Firewood-Gatherer or Leñatero (*Anumbius annumbi*) is an 8-inch (20-centimeter) bird, light brown above with a rufescent forehead, white eyebrows, white throat outlined by small black spots, and buffy breast and abdomen. In northeastern Argentina, Uruguay, Paraguay, and southeastern Brazil, it inhabits open woodlands and weedy fields, where it walks over the ground gathering its insect food. With interlaced twigs that are often thorny, this ovenbird builds a globular or cylindrical nest from 6 to 65 feet (2 to 20 meters) up in a tree; the nest may be used for at least two years and become 43 inches long by 18 inches wide (109 by 46 centimeters). The massive structure has an obliquely descending entrance tube which is sometimes spiral and, in the larger nests, up to 31 inches (79 centimeters) long. The chamber in the lower part of the nest is thickly lined with feathers, seed down, wool, rabbit skin, snakeskin, and the like. Here three to five white eggs are incubated by both sexes for sixteen or seventeen days, and the nestlings are fed for twenty-one or twenty-two days. Fledglings are led back to sleep in the nest, which during the winter continues to be used as a dormitory, at least by the adults. More frequently than among the Rufous Horneros, the young Firewood-Gatherers help their parents build nests, sometimes starting to carry twigs when only forty days of age. However, their service as helpers is brief, for most are expelled from the parental nest sites before they are two months old.

An equally industrious builder is the smaller, brownish, wrenlike Plain-fronted, or Rufous-fronted, Thornbird (*Phacellodomus rufifrons*), which inhabits three widely separated areas in northern Venezuela, in northern Peru, and from the hump of Brazil to Bolivia and northwestern Argentina. In dense thickets and weedy fields, thornbirds rummage for insects and spiders amid fallen leaves, in piles of brush, or in vine tangles at no great height. Although they are secretive while they forage, for their nests they prefer isolated trees, where their great structures of interlaced twigs hang in plain view from about 7 to 75 feet (2 to 23 meters) above the ground. In Venezuela the nests are conspicuous along busy highways and in the scattered trees of the *llanos;* the thornbirds' habit of plucking off surrounding foliage increases their exposure. These pendent structures may become 7 feet high and 15 to 18 inches in diameter (2 meters by 38 to 46 centimeters) and contain up to eight or nine chambers, usually one above another, rarely two side by side. Most are smaller, but the smallest occupied nests have at least two rooms. These never intercommunicate; each has its separate entrance from the outside. The chambers are lined with the most diverse materials, including fibrous bark, grass, feathers, and, near human habitations, scraps of cellophane, colorful candy wrappers, tinfoil, paper, and other trash. In the lowest chamber, which would be least accessible to a snake or other wingless predator advancing along the branch from which the structure hangs, the female lays three white eggs, which she and her mate incubate about equally for sixteen or seventeen days.

Fed by both parents with larval, pupal, and mature insects, rarely a spider, the young thornbirds fly from the nest when twenty-one or twenty-two days old. In the evening, their parents lead them back to sleep in the nest, which (if no mishap befalls it) will be their nightly lodging until they are over one year old. Nests with three to six self-supporting occupants are not rare. Including two parents, four older young, and two fledglings, one nest sheltered eight sleepers. The young of one or two annual broods commonly continue to forage and lodge with their parents until well into the following breeding season.

The Plain-fronted Thornbirds' multichambered nests remind one of the apartment nests of Palm Chats (*Dulus dominicus*), Monk Parakeets (*Myiopsitta monachus*), and Sociable Weavers (*Philetairus socius*), and one expects that, like these others, they would be occupied by several breeding pairs—especially when half a dozen grown birds are seen near them. Nevertheless, thornbirds are strongly territorial; none of the more than twenty nests that I watched in Venezuela was occupied by more than one breeding pair, often with their full-grown, nonbreeding offspring. Contrary to expectations, these unmated birds appear not to feed the nestlings or fledglings with whom they are so closely associated. However, they help build and maintain the structures in which they lodge. In nearly six hours of watching, Betsy Thomas saw the male of one pair bring 112 sticks, the female 69, while two of their three young, about four months old, contributed 17 small pieces, chiefly for the nest lining.

Likewise, interspecific helpers have rarely been found among ovenbirds, probably because many are so retiring and difficult to watch. A pair of White-winged Cinclodes (*Cinclodes atacamensis*) helped a pair of Gray-flanked Cinclodes (*C. oustaleti*) feed their feathered nestlings in a burrow 24 inches (61 centimeters) long.

REFERENCES: Fraga 1979, 1980; Hudson 1920; Narosky et al. 1983; Skutch 1969a; Thomas 1983.

21.

AMERICAN FLYCATCHERS

Tyrannidae

The 376 species of flycatchers spread widely over the continents and islands of the western hemisphere, where they are most abundant in the tropics. Those that nest where winter is severe migrate to avoid it. In length they range from 3 to 16 inches (8 to 41 centimeters), but only those with very long tails exceed 11 inches (28 centimeters), and the majority are small. Most are plainly attired in shades of olive, gray, and brown; many have yellow breasts. A few are largely black, white, or both together; one is vermilion. Except for a few that forage over the ground, they are arboreal birds that with short, often flattish bills catch insects in the air or pluck them from foliage. Some eat berries and other fruits.

Most of the few species that have been carefully studied breed in monogamous pairs; the female builds the nest, incubates the eggs, and broods and feeds the young, sometimes assisted by her mate in building, nearly always in feeding, but scarcely ever in incubating. In a substantial minority of the species for which we have information, the sexes form no lasting attachment and the female nests quite alone. Flycatchers' constructions are exceedingly diverse and include open cups, roofed nests with a side entrance, and a great variety of pensile structures. The two to six eggs are plain white or variously tinted, spotted, and streaked. Of the many American flycatchers, only two species are known to breed cooperatively, at least occasionally.

Although these two cooperative species are not closely related, they look alike. Both have black or blackish caps, separated by prominent white eyebrow stripes from the black cheeks. The upperparts are brownish olive; the throat is white; and the remaining underparts are bright yellow. With minor variations, this color pattern is shared by a number of flycatchers of several genera. The 7-inch (18-centimeter) Rusty-margined Flycatcher (*Myiozetetes cayanensis*) has an orange crown patch, rarely revealed unless the bird is excited, and rufous-margined wing feathers. It is widely distributed from Panama to western Ecuador, Bolivia, and southern Brazil. The slightly smaller White-bearded Flycatcher (*Conopias inornata*) lacks the bright crown patch widespread in flycatchers and has no rufous on the wings. A bird of open country, it is found only in central Venezuela.

On the Venezuelan *llanos*, wide grassy plains with scattered palms and other trees, swept by drying winds for six hot, rainless months and flooded by tropical downpours for the remainder of the year, Betsy Thomas studied White-bearded Flycatchers. The nest, a neat, shallow, open cup, composed of a variety of vegetable materials with lichens attached to the outside, blends with the bark of the small tree where it is built on a horizontal fork. Two white eggs, sparsely dotted and blotched with chestnut, are laid, with an interval of two days, from early April to mid July. Of three carefully watched nests, one belonged to an unaided pair and each of the other two was attended by a trio. At each of these nests with a helper, all three adults built, fed the nestlings, and defended the territory. Two of them shared incubation and brooding rather equally; the third did so occasionally and apparently not very effectively. About one-quarter of all the nestlings' meals were brought by helpers, who also carried away their droppings. When, after seventeen or eighteen days of incubation and a nestling period of eighteen days, the young left the nest, the helper was not seen to feed them, although it assisted in the defense of the fledglings and the territory. One young flycatcher continued to follow and beg from its mother until it was three months old, although at half this age it began to catch insects by itself.

These White-bearded Flycatchers behaved like typical cooperative breeders. Parents were never antagonistic to their helper. They engaged in duets, accompanied by up-and-down movements of their bodies and rhythmic flapping of their wings; sometimes one joined in a duet while sitting in the nest. Juveniles preened each other. After the young flew, all members of the social group foraged and roosted close together. The male's participation in incubation and brooding sets White-bearded Flycatchers apart from all the other flycatchers that I know.

Using grasses and other vegetable materials, Rusty-margined Flycatchers build a well-roofed nest with a side entrance, much bulkier than the White-bearded's compact cup. The only nest of this widespread bird of which cooperative breeding has been reported was situated on a branch of a dead tree that had fallen into Gatún Lake in the former Panama Canal Zone. Here, in watches totaling six hours, Robert Ricklefs saw four adults, alike in appearance, feed three half-grown young 211 times. These attendants had a strong tendency to approach the nest together rather than independently and to feed the nestlings in the presence of one or more of their associates. While delivering food at a covered nest, the feeder's back is usually exposed but its head is inside, so that it cannot see the approach of an enemy. Ricklefs suggested that the behavior of these Rusty-margined Flycatchers might be an adaptation to reduce the risk of predation. While one member of the

White-bearded Flycatcher

group feeds the nestlings, others serve as sentinels to warn of approaching danger.

Once at nightfall I watched three Rusty-margined Flycatchers roosting on a low shrub in an exposed situation. The three were in close contact, their yellow breasts all turned the same way. Fledgling flycatchers of a number of other species sleep pressed together, while adults of these same species roost a short distance apart. The three Rusty-margined Flycatchers appeared to be adults, preparing to sleep in a manner widespread among birds that practice cooperative breeding. Probably these flycatchers do so not infrequently.

Flycatchers occasionally help birds of other species. An Eastern Phoebe (*Sayornis phoebe*), whose first brood was becoming independent, brought food to nestling Tree Swallows (*Tachycineta bicolor*), continuing this for about a week, while the parent swallows tried to drive her away. An Eastern Kingbird (*Tyrannus tyrannus*) fed nestling Baltimore, or Northern, Orioles (*Icterus galbula*), at first against parental opposition. After a while, parents and helper became reconciled; and the kingbird continued to feed the young orioles after they left the nest. By their bold attacks on predators, especially those that fly, nesting kingbirds and others of the larger flycatchers help protect the eggs and young of their smaller feathered neighbors. Additional cases of interspecific helping found in the literature by Marilyn Shy include a Scissor-tailed Flycatcher (*Tyrannus forficatus*) who fed Common Grackles (*Quiscalus quiscula*) in a neighboring nest, an Eastern Wood-Pewee (*Contopus virens*) who gave food to orphaned nestling Eastern Kingbirds, and a Least Flycatcher (*Empidonax minimus*) who nourished Chipping Sparrows (*Spizella passerina*) in a nest near its own.

REFERENCES: Bragg 1968; Deck 1945; Ricklefs 1980; Shy 1982; Thomas 1979.

22.

COTINGAS

Cotingidae

The cotingas are an extraordinary family of controversial limits. The recent tendency has been to remove the tityras, becards, and mourners, long held to be members in good standing. Even with this reduction, the cotinga family is an extremely varied assemblage of about sixty-five species, ranging in size from 3.5 to 18 inches (9 to 46 centimeters), thus containing some of the smallest as well as the largest passerine birds. All are arboreal birds confined to the wooded regions of continental tropical America. Males may be wholly white, wholly black, or largely red, orange, green and yellow, shining blue and purple, rufous, or gray. Some males have curious wattles. Females of the more ornate males are more plainly attired.

Most cotingas prefer a mixed diet of fruits and insects; a few are wholly frugivorous. Some have stentorian voices, freely used; others are strangely silent; few, if any, are truly songful. Their nests, in trees or shrubs or attached to rocks, may be shallow saucers or deeper cups or platforms barely large enough to hold a single egg. In many species, the female alone builds, incubates the one to three buff or olive, spotted or blotched eggs, and feeds the nestlings, while the male remains aloof and calls or indulges in fantastic displays. In most of these features, members of the reduced family differ strongly from the excluded tityras and becards.

Unique among the better-known members of the family is the Purple-throated Fruitcrow (*Querula purpurata*), which lives in upper levels of rain forests from Costa Rica to western Ecuador and, on the other side of the Andes, in Venezuela, the Guianas, and over much of Amazonia to northern Bolivia and central Brazil. The male, about 11 inches (28 centimeters) long, is black with an expansible crimson gorget; the slightly smaller female is everywhere black. In groups of three or four to six or eight, these cotingas seek both fruits and insects, which they usually seize in flight, in typical cotinga fashion. They reveal their presence in the leafy treetops by delightful liquid calls delivered with closed bills and by less pleasant throaty notes. Flock members perch side by side, almost touching, preen each other, and roost in close contact on a horizontal limb. They are almost never aggressive toward their companions.

Nests of Purple-throated Fruitcrows have rarely been reported. The only detailed study of their breeding was made by David Snow in the Kanuku Mountains of southern Guyana, where two nests were found. They were approximately 35 and 70 feet (11 and 21 meters) up in trees whose crowns did not touch those of other trees, so that flightless animals could reach them only by climbing the trunks from the ground—a provision apparently important for the nests' safety. The open cup of twigs and

Purple-throated Fruitcrow

dry fruiting panicles was so loosely constructed that the egg was visible through chinks in the bottom. At the two nests watched by Snow, only the females were seen to bring materials; but the males, obviously interested, occasionally sat in the nests and helped shape them. In Panama, however, Hazel Ellis, watching earlier stages of construction, saw the male take many pieces to the nest, while the female shaped it. Once the male and female carried a yard-long piece of vine between them, each holding one end. Nearly a month elapsed between the discovery of Snow's first nest, when construction was far advanced, and the laying of the single deep olive egg, thickly marked with blackish brown. At the one successful nest, only the female incubated, sitting for about two-thirds of the daytime.

During building and the twenty-five-day incubation period, this female fruitcrow was closely associated with two males and another grown bird of undetermined sex, who was molting wing feathers and possibly for this reason was less helpful. After the downy nestling hatched, all four adults brought it food, chiefly insects, with a few fruits after it was older. The primary male, who was probably the nestling's father, fed most often, followed by the second male, then by the mother and the bird of undetermined sex, who contributed little. Between them, the four attendants offered the single nestling much more than it could eat. All spent much time near the nest and mobbed intruders, chiefly toucans, jays, and hawks, by swooping within a few inches of them on swift aerial dives, pursuing them closely, and harrying them after they took wing. With efficient guarding by these four spirited attendants, the young fruitcrow remained safely in its exposed nest until it flew at the age of thirty-two or thirty-three days—twice the nestling period of a thrush (*Turdus*) of similar size.

References: Ellis 1952; D. W. Snow 1971, 1982.

23.

NEW ZEALAND WRENS

Acanthisittidae

Four species (one recently extinct) comprise this family of tiny insectivorous birds, confined to the archipelago for which it is named, unrelated to the wrens of the family Troglodytidae. Most abundant and familiar of the three extant species is the Rifleman (*Acanthisitta chloris*), an almost tailless birdling about 3 inches (8 centimeters) long. The sexes differ in both size and plumage, the females being larger and more highly colored—the reverse of the situation usual in passerines that show sexual dimorphism. Largely olive-green above, they have prominent white eyebrows and whitish underparts washed with yellow on the flanks. The short, sharp bill is slightly upturned. The Rifleman remains abundant in the beech (*Nothofagus*) forests of both main islands and is becoming adapted to cultivated areas. It flies weakly with short, rounded wings, and with strong legs it spirals up trunks and stout branches, plucking insects and spiders from crevices in the bark or amid mosses and lichens. From high on one tree it flies to near the foot of another to repeat the process. Its high-pitched, rapidly repeated *zee* or *zsit* appears to be responsible for its name.

In a hole in a trunk or branch, the Rifleman constructs, of loosely woven roots, leaves, mosses, and ferns, a domed nest with a side entrance, which it lines with feathers. The male not only does most of the building but also feeds his

Rifleman

mate generously while she forms her plain white eggs. Each weighs about 20 percent as much as she does, so that her clutch of five, laid at intervals of two days, equals her own weight. The male also undertakes most of the daytime incubation during a period of nineteen to twenty-one days, long for so small a bird. At first he is the nestlings' chief provider, but after their first week from one to three helpers arrive, bring them food, remove their droppings, and aid in defending them from predators. These assistants include adults and juveniles of both sexes, but most are unpaired males, who are not always related to the parents. Although some regularly attend a single nest, others bring food sporadically to several nests. When, at from twenty-three to twenty-five days of age, the young leave the nest, they weigh more than the adults, but those fed by both parents and helpers do not become heavier than those nourished by the parents alone.

While their first brood is still in the nest, parent Riflemen have usually started to build a nest which is smaller, more loosely constructed, and often unlined. Probably because he is still feeding young of the first brood, the male parent does not bring food to his mate while she lays her second clutch, which averages one egg smaller than the first. As the season advances, most adult male helpers pair with young females from the brood that they have earlier attended—one reward for their services.

Reference: Sherley 1985.

24.

SWALLOWS

Hirundinidae

Swallows should need no introduction to readers of this book. Their seventy-four species are distributed all over the Earth, except for the polar regions, New Zealand, and some oceanic islands. They pass much of the day tracing erratic courses above our heads while they catch flying insects, almost their only food, over towns and villages as well as open fields and waters. When tired, they rest conspicuously, strung out on service wires or on bare branches instead of hiding amid foliage, as many small birds do. They migrate in large flocks by day, instead of by night, the more usual way. Some nest in colonies, often on our houses or other buildings, where their songs, pleasant but not brilliant, endear them to us. We would miss them if the pesticides that we spread ever more recklessly destroyed all the insects on which they subsist.

One of the most social and best-studied of the swallows is the steel blue–and–white House Martin (*Delichon urbica*), which breeds over Eurasia from the British Isles to Japan, migrates in winter to southern Africa, and has won a place on the North American list by its accidental occurrence in Alaska. It nests in colonies that sometimes contain several hundred pairs, which originally plastered their nests of mud on vertical cliffs beneath an overhang that protected them from dissolving rain but now more frequently construct them on our dwellings, beneath eaves or in other sheltered places. With pellets of mud or wet clay, both sexes build a roughly globular urn with a narrow orifice at the top, then line it with feathers or other soft materials. The martins gather their mud in social parties, and occasionally a building pair is helped by its neighbors. Fourteen individuals assisted in the construction of one nest, and when this was finished, they helped with others. Esko Lind, who for years studied House Martins in Sweden, watched four or five birds build the last nest in a colony, completing it in six days instead of the usual ten—as though to help the belated pair make up for lost time! When a nestling hung precariously to the side of its nest, unable to regain the interior, several martins built a sort of antechamber to support it. A single male and female alternately incubate the eggs, with very frequent changeovers and apparently no help from other martins. At night, both sleep in the nest with eggs and nestlings.

Auxiliaries, up to four or five in number, may help the parents feed a brood. When a flightless nestling fell, from three to five adults brought food to it on the ground for four days. Juveniles sometimes help feed siblings of later broods, and Lind repeatedly watched young martins take food to nestlings in neighboring nests. House Martins usually remain in their snug urns for twenty-four or twenty-five days, sometimes as long as twenty-eight days, and do not venture into the outside world without prompting. As much as ten days before the young birds are ready to fly, adults try to lure them from the nest by flying slowly with rapidly fluttering wings in front of it, feigned feeding, and special calls. Up to twenty individuals may participate in this luring ceremony. When sufficiently mature, the fledglings respond to these solicitations, take wing, and follow the white rumps of their

House Martin

elders. At this critical moment in the lives of young martins, the cooperation of colony members is important, for the parents themselves can guide no more than two fledglings from the nest at a time. The young return to sleep with their parents in the feather-lined urn of clay. In Switzerland, both parents and eight juveniles from two earlier broods slept with three nestlings of the third brood, a total of thirteen martins in the crowded nest.

Also closely associated with humans is the Purple Martin (*Progne subis*), which breeds over most of the temperate zone in North America and winters in tropical South America. It appears to prefer the birdhouses and gourds that people provide for its nests to the holes in trees that it occupied before Europeans arrived. When William Southern carried nesting females from a colony of about eighty pairs in Michigan to test their ability to find their way home, neighboring martins, usually females but occasionally adult or immature males, fed the absent parents' nestlings at ten nests. Sometimes they continued this service after the parents' return. The mother of a brood in an adjoining compartment in the birdhouse fed the absent mother's nestlings as well as her own, while the nestlings' father accepted this assistance without protest. At colonies in Texas studied by Charles Brown, nine males disappeared over the years, leaving their mates with nestlings. Two of these widowed females were joined by males who helped them feed their broods. Although more unmated males were available, and males sometimes peered into the nests with fatherless young, the other seven broods were successfully reared by their unaided mothers.

Other cases of helpers among Purple Martins include a hand-reared female who, at the age of fifty-four days, tried to brood nestlings and soon began to bring insects to them, continuing this actively. When four nestling European Starlings (*Sturnus vulgaris*) moved from their compartment in a birdhouse to the adjoining compartment where martins nested, both parent martins fed them repeatedly.

The Brown-chested Martin (*Phaeoprogne tapera*) is widespread in South America, from the Caribbean coast to central Argentina. Like other swallows that breed in the temperate zones, those that nest farthest south are migratory. In Argentina, where this martin occupies nests of Rufous Horneros (*Furnarius rufus*), Rosendo Fraga watched nine families of fledglings, only one of which was attended by more than the two parents. The exceptional group was regularly fed by four or five adults, probably parents with siblings from a previous brood.

The Barn Swallow (*Hirundo rustica*), known in Britain simply as the Swallow, is one of the most cosmopolitan of birds, for in addition to nesting over most of the north temperate zone it winters in much of the southern hemisphere. While the House Martin prefers human habitations for its closed nests, Barn Swallows choose the interior of more open structures, such as barns and other outbuildings, for their open cups of mud lined with feathers. In both Europe and North America, the young of early broods have frequently been watched while they fed their parents' subsequent brood. Sometimes they help build the nest for this brood; in one instance they engaged in this work about a week after they first flew.

Less frequently reported is the presence of adult auxiliaries among the attendants at a nest. In Germany, two pairs of Barn Swallows nested in a barn. After one male vanished, the surviving male helped feed the young in both nests. Later, he apparently fathered a second brood with each female parent—a rare case of bigamy in this species. At a colony of banded swallows in Ohio, helpers aided substantially in feeding the young at eight of twenty-one nests, accounting for 20 to 28.5 percent of all feeding visits at some of these nests. Contrary to the expectation based on earlier reports, these helpers were nearly all adults; at only one of the eight nests were juveniles seen to feed younger siblings. In addition to colony members who served as helpers, a flock of about eighteen unidentified Barn Swallows arrived from a distance and contributed more than 19 percent of the meals brought to the young.

Across Canada to the northern limit of trees and southward through much of the United States, the Tree Swallow (*Tachycineta bicolor*) nests in hollow trees and birdhouses. More than two attendants at a nest are not uncommon. The nestlings in one box were fed by four to six adults, of whom at least three were males and at least one a female. In another locality, at least three broods each had an extra attendant, a female whose own young had died. In the related Violet-green Swallow (*T. thalassina*), which breeds on the western side of North America from central Alaska to the highlands of Mexico, two or three females sometimes feed the young in a single nest. Over almost the whole of temperate and subarctic North America, the Cliff Swallow (*Hirundo pyrrhonota*) fastens its small-mouthed urns of hardened mud or clay to the faces of cliffs and buildings in large, crowded colonies. Occasionally three individuals build together and incubate the eggs by turns.

Over much of the north temperate zone, from Alaska to Siberia the long way around, Bank Swallows (*Riparia riparia*), known in the Old World as Sand Martins, burrow into exposed banks of rivers, seaside cliffs, or sand pits to build their nests and lay their four to six white eggs. In suitable localities, their colonies may contain hundreds of pairs. In England, but apparently not in North America, the excavation of the burrows is the climax of an elaborate aerial display in which many birds participate. The performance, as described by R. A. O. Hickling, begins when dispersed swallows feeding high overhead gather quickly into a flock, which becomes ever more compact as it spirals downward to the face of the cliff, where, after hovering briefly, the birds alight at the entrances of newly begun holes. At one display, four to six birds may cooperate at one hole, up to three entering to dig together, while others await their turns beside the burrow's mouth. At the next display, the swallows may excavate at quite different tunnels. The same individual may move from hole to hole, digging at several in succession and throwing out jets of sand. The number of burrows is at first approximately equal to the number of participating swallows, but not all are occupied for nesting. After cooperating to excavate their burrows, Bank Swallows build their slight nests, lay their eggs, incubate, and rear their young in monogamous pairs.

Despite the frequent occurrence of helpers in several species and at various stages of the reproductive cycle, no swallow for which adequate information is available qualifies as a typical or advanced cooperative breeder. Although many are highly gregarious, they are not organized into the cohesive, structured groups that we expect in cooperative breeders. Except in a few activities, such as leading fledgling House Martins from their nests and burrow digging by British Bank Swallows, their helpfulness appears to be sporadic rather than habitual. Unlike many cooperative breeders that rest in closest contact, when swallows perch in a row, as on an overhead wire, they maintain their individual distance, often separated from their nearest neighbors by about the length of their own bodies. They preen themselves, not their companions. To be sure, in inclement weather House Martins and others huddle together in compact masses of many individuals, their wings sticking out in all directions, but this is to preserve warmth rather than because they love to be so close to companions other than their immediate families. As Lind has illustrated, House Martins often make long extensions from their nests to prevent other pairs from building too close to them. Moreover, the swallows about which we have most information are both colonial and long-distance migrants, whereas cooperative breeders are usually group territorial and permanently resident, so that the young can remain with their parents from year to year. Helpers have apparently not been reported for the few permanently resident tropical species of swallows for which we have information.

REFERENCES: Bent 1942; C. R. Brown 1977, 1983; Fraga 1979; Hickling 1959; Lind 1960, 1964; Mohr 1958; Myers and Waller 1977; Richmond 1953; W. E. Southern 1959, 1968; Van Velzen 1960; Witherby et al. 1938.

25.

CROWS AND JAYS

Corvidae

The crows, ravens, rooks, jackdaws, jays, choughs, magpies, nutcrackers, and their allies constitute a family of 113 species, distributed over the whole Earth except southern South America, New Zealand, some oceanic islands, and Antarctica. The number of names they have received from people other than ornithologists is evidence of our long familiarity with these abundant, aggressive birds. In size they range from the 8-inch (20-centimeter) Dwarf Jay (*Cyanolyca nana*) of Mexico to the 26-inch (66-centimeter) Common Raven (*Corvus corax*), the largest passerine bird. In contrast to the somber black of crows and ravens, many members of the family are bright with blue, green, or yellow, and between these extremes of coloration are plumages that are largely gray, brown, or boldly black and white. Crests and notably long tails adorn a few species. Almost omnivorous, crows and jays consume a great variety of animal and vegetable foods, including grains, nuts, and soft fruits. Many species store food for future use. Apparently none is a fine songster—their voices are mostly loud and harsh—but in solitude some sing pleasing medleys in subdued tones. Some display a certain ability to mimic, especially in captivity.

Most crows and jays build open cups of coarse sticks, lined with softer materials, in trees or shrubs, in crannies, or on sheltered ledges of cliffs. A few breed in holes in trees or buildings, and magpies construct roofed nests with a side entrance. The two to eight, rarely more, eggs are greenish, bluish green, gray, buffy, or olive, usually spotted or blotched with shades of brown and pale lilac. Except in nutcrackers, only the female regularly incubates, fed by her mate and often by helpers. Both sexes nourish the young.

Let us begin with the first advanced cooperative breeder to be studied in some detail, the big 17-inch (43-centimeter) Brown Jay (*Cyanocorax morio*), which from extreme southern Texas ranges through eastern Mexico and Caribbean Central America to western Panama, from the coast ascending as high as 8,300 feet (2,530 meters) in Costa Rica. In northern Mexico, this exceptionally plain jay is almost wholly dark brown; in Central America, it has broadly white-tipped outer tail feathers and a whitish abdomen; in southern Mexico, the two color phases intermingle. The yellow bills, eye rings, legs, and feet of juveniles darken with the years until they become wholly black. The great diversity of color patterns on the pied bills of jays in intermediate stages facilitates individual recognition—a boon to the student of their social habits.

Although in plumage so different from other New World jays, the brown bird's behavior is too jaylike to leave one long in doubt as to its classification. Avoiding the tall rain forests that not long ago covered most of the Caribbean slope of Central America, it frequents the great banana plantations that have replaced the forests, bushy pastures with scattered, vine-draped trees, light second-growth woods, and shady margins of the wider streams. In the highlands of Costa Rica it lives amid much lighter vegetation, including thinly shaded coffee plantations, pastures, groves of small trees, and wooded ravines. Among the first birds to become active at dawn, it is also one of the last to become silent at nightfall. With a volley of *chaa*'s, it protests any intrusion into its haunts, then flees with a headlong, rocking flight. A foraging flock of five to ten Brown Jays is extremely difficult to surprise and to watch. They eat many fruits and are fond of the nectar of banana flowers, which they sometimes pluck and hold beneath a foot while they extract the sweet liquid. They probe curled dead leaves for insects and spiders, catch dragonflies and lizards. When foraging on the ground, they fly up the moment they find themselves observed. In humid lowlands where food is abundant, I never saw them take eggs or nestlings, but amid the sparser vegetation of deforested highlands they pillage many a nest.

Like a number of other cooperative breeders, immature Brown Jays indulge in an activity that appears to be play. Perching in front of a companion, a young jay stretches up its legs and makes feints with its bill toward the other, now here, now there, bobbing up and down, twisting and turning from side to side in a spirited manner, until it resembles a feathered clown. The passive bird directs its beak toward the actor and erects the feathers of its head and upstretched neck, looking very bizarre as it silently endures the mock attack. Later, the roles of the two participants in this play may be reversed.

In the Caribbean lowlands of northern Central America, Brown Jays may begin to build their nests in February, but few have eggs before April. Their bulky structures are situated in the crown of a banana plant or in a tree, from about 12 to 75 feet (4 to 23 meters) above the ground. Male and female share the strenuous work of breaking coarse twigs from trees for the foundation and pulling fibrous roots from bare ground for the lining. Each sits in the nest to shape it. Rarely a young, unmated bird brings a stick. When not actively building, one member of the pair, probably the female, spends much time sitting in or near the nest and repeating a plaintive *pee-ah,* the nest call.

On consecutive days, the female lays two or three bluish gray eggs thickly speckled with brown. During the incubation period of eighteen days, she sits very constantly, often from two to nearly four hours at a stretch. When hungry, she repeats the loud nest call, which sounds like a complaint, and draws not only her black-billed mate but also helpers with pied, black-and-yellow bills. At one nest at least three attendants brought food to her. Sometimes, after her mate has fed her, she flies off to forage for herself, leaving him standing guard beside the nest until she returns, rarely more than a quarter of an hour later. If her attendants have neglected her until she becomes very hungry, she may leave the nest unguarded while she seeks food, but she returns more promptly.

Newly hatched Brown Jays have yellow skins, sometimes tinged with olive, wholly devoid of down. Their eyes are tightly closed, their bills and feet are yellow, and the interior of their mouths is red. Their feathers sprout so slowly that they are mostly naked until about sixteen days old, but by their twentieth day they are fairly well clothed. Unless disturbed, they do not abandon their nests before they are twenty-three or twenty-four days old, when they resemble their parents in plumage but still have yellow bills, eye rings, and feet. Until they are feathered, they are brooded by day by their mother and, on rare occasions, by their father. They are fed not only by their parents but also by helpers, of which, at four nests, I distinguished by their bills one, two, three, and five.

At the nest with five auxiliaries, each, after feeding the young, remained on guard until another arrived with food, so that while still naked the nestlings were rarely left alone. Instead of giving food directly to the nestlings, one attendant often passed it to another for delivery. A helper might entrust food to a parent or another helper; a parent might give it either to the other parent or to a helper. Rarely, a jay with a largely yellow bill carried away the food it had received from another instead of delivering it to a nestling. Sometimes, too, an attendant failed to remain on guard after feeding the nestlings. Meanwhile the previous sentinel, doubtless expecting to be replaced, had flown away, so that the nest remained unguarded until another jay arrived. In 13.5 hours of watching at this busy nest, I saw the three naked nestlings fed by the seven attendants a total of 91 times, at the rate of 2.2 times per nestling per hour. After the nestlings were feathered, they were fed by the same purveyors 30 times in 3 hours, at the rate of 3.3 times per nestling per hour. Now the attendants stayed at the nest only long enough to deliver food and remove droppings. Perhaps because they were no longer well guarded, the nestlings were taken by a predator.

In addition to feeding the young jays, removing their droppings, and watching over them, the helpers tried to protect them. Probably because they were younger and less prudent, their zeal as defenders exceeded that of the parents. At a nest in a willow tree to which I could climb, a pied-billed helper would dart within a few inches of my head or perch a yard above it and vociferate. Lacking the courage to attack me directly, as some much smaller birds have done, it relieved its feelings by pecking vigorously on a branch or tearing a nearby banana leaf to shreds. After the parents and their two other assistants, tiring of their demonstrations, had retired to a safe distance, this zealous guardian remained close by until I had finished my examination of the nestlings and descended to the ground.

Years later, Marcy Lawton and Carlos Guindon studied Brown Jays in trees growing isolated amid pastures at Monteverde, around 5,000 feet (1,525 meters) above sea level in the Cordillera de Tilarán in northwestern Costa Rica. Here the jays lived in groups of six to ten individuals on overlapping home ranges of 25 to 50 acres (10 to 20 hectares). As elsewhere, each group had a single nest that was attended by both breeding and nonbreeding members. Although in the Caribbean lowlands I have never found more than three eggs or young in a nest or seen any indication of more than one breeding pair in a group, nests at Monteverde held from three to seven eggs, suggesting, with other indications, that two or more females sometimes bred together. Classifying the six to ten attendants at these nests as old, intermediate, and young, Lawton and Guindon found that the success of nests depended upon the age and experience of the helpers more than upon their number. On the whole, these jays nested quite successfully, producing thirty-four fledglings from fifty eggs (68 percent).

This study demonstrated how young Brown Jays became more proficient as they gained experience while attending a single brood. At first they frequently made mistakes, such as offering very young nestlings items too big for them, eating what they had apparently brought for the young and then approaching the nest with empty bills, or, after alighting on the nest and uttering the feeding note, being distracted by some disturbance in the neighborhood and flying away with undelivered food. The youngest jays made such blunders more frequently than those of intermediate age, and the latter blundered more often than the oldest birds. Although the performance of young jays improved markedly while they attended a brood, that of older individuals, who were more efficient at the beginning, hardly changed.

One of the most elegant of New World jays is the Green (*Cyanocorax yncas*), which is found from southern Texas through Mexico to Honduras and, after a wide hiatus in its range, in South America from northern Colombia and Venezuela to eastern Peru and northern Bolivia. Its back, wings, and central tail feathers are green, its crown and nape white or violet-blue according to the race, and its throat black. Its outer tail feathers and more posterior underparts are bright yellow in South America, but in the north its breast and belly are pale green.

For a year, Humberto Alvarez followed four groups of these handsome jays amid pastures, cultivated fields, tree plantations, and patches of second-growth woods at an altitude of about 8,200 feet (2,450 meters) in the Central Andes of Colombia near Medellín. Each group was a stable association of from five to eight adult and immature jays, who remained on their territories throughout the year. They lived in amity, resting in contact, preening each other, and rallying to the spirited defense of any group member threatened by a hawk.

Nesting started shortly after the return of the rains in March and continued until mid June, the peak of the wet season. After an interval of courtship, which included nuptial feeding and allopreening while they perched in contact, the single mated pair of each group proceeded to build a nest, from 16 to 33 feet (5 to 10 meters) up in a tree, with occasional assistance from other group members. On consecutive days, each female Green Jay laid four eggs, similar to those of the Brown Jay, which were sometimes visible through interstices in the bottom of the loosely constructed cup. During the incubation period of seventeen to eighteen days, the mated females sat closely, keeping their eggs covered for 82 percent of the daytime at one nest and for 85 percent at another. Each was fed rather infrequently, chiefly by her mate, either while she covered the eggs or when she

Brown Jays, showing adult (left) and immature (center and right) bill coloration

left to join the flock. After giving her food, the male might follow her back to the nest. Other flock members fed her only during the last three days of incubation. Once, during a breeding female's absence, an immature jay sat in the nest for three minutes, until the other returned and pushed it aside. All group members came at intervals to inspect the nest, whether the female was incubating or absent.

After the eggs hatched, the auxiliaries became more active at the nests. From one to five of these helpers aided the parents in feeding the three or four young in each of three groups. At some nests they contributed more than half of the meals; mostly they brought food more often than the nestlings' mother, who was the least active feeder. However, she did all the brooding, except for very brief, incompetent attempts to cover the nestlings by their father or a helper. At first, an attendant arriving with food gave part of it to the brooding mother or any other group member who happened to be present, then each fed the young. This readiness to share decreased rapidly after the nestlings were three days old. Then, when the food bringer did not comply with the solicitation of a jay already present, the latter might try to take it forcibly from the newcomer's bill, and both tugged at the item until one succeeded in wresting it from the other and quickly passed it to a nestling. The father, who brought the biggest meals, was always more willing to surrender part of them to a helper for delivery to the nestlings. After feeding, an attendant often remained to guard the nest in the mother's absence but rarely lingered after her return.

If undisturbed, the young Green Jays left the nest when from nineteen to twenty-two days old and promptly hid amid dense vegetation, where they continued to be fed by both parents and helpers for at least twenty days longer. Their first weeks out of the nest are a perilous interval for fledglings—many are lost. During these weeks the value of attendants additional to the parents was most clearly demonstrated, for while some mobbed and threatened a potential predator, others led the young jays away and fed them at intervals between harassing the enemy. Of fourteen young fledged by four groups with a total of eight helpers, 10 (71 percent) were still alive when Alvarez ended his study from three and a half to five months after they left the nest—a good record.

Amid the deciduous woods and arid scrub of the Pacific side of Middle America, from the Mexican state of Colima southward to the Gulf of Nicoya in Costa Rica, and in certain arid, mountain-rimmed valleys in the Caribbean drainage, one of the first birds likely to attract the traveler's attention is the handsome White-throated Magpie-Jay (*Calocitta formosa*). The long tail of this 21-inch (53-centimeter) bird and its high crest of waving, forwardly curved black feathers distinguish it from every other New World jay except its close relative, the even longer-tailed Black-throated Magpie-Jay (*C. colliei*), which replaces it in northern Mexico. The white-throated jay's upperparts, wings, and central tail feathers are blue; its face, underparts, and outer tail feathers are white; and a narrow black band crosses its breast. Small, straggling flocks of these big jays follow the human intruder into their domain, abusing the alien with volleys of excessively harsh notes and warning the whole feathered community of a foreign presence. On occasion they utter mellow, liquid calls; and while resting inconspicuously amid foliage they amuse themselves with a medley of queer notes.

In different parts of their range these jays can be found, from December until July, sitting high in a tree upon three or four gray eggs finely and densely flecked all over with brown, in the jay's usual

Top to bottom: Gray Jay, Yucatán Jay, Tufted Jay, White-throated Magpie-Jay, Scrub Jay, Green Jay

nest of coarse sticks lined with wiry roots and fibrous materials. This jay has never, to my knowledge, received the careful study that it merits; but long ago I passed most of Christmas Day and part of the following day watching such a nest, inaccessibly situated nearly 100 feet (30 meters) up in a tall, clean-boled tree on a Guatemalan coffee plantation. Only the female, whom I could distinguish from her associates by her loosely spreading crest, incubated the eggs, for 86 percent of the nearly fourteen hours that I watched.

Much of the time the incubating jay was hungry, as I inferred from the frequent repetition of cries which, although loud and harsh, were nevertheless somewhat pleading. In response to these supplications, she was fed forty-seven times certainly by two, and probably three or more, of the adult jays who were often near her nest. Once she was fed five times in four minutes. At the approach of an attendant with food, she spread her wings over the sides of her nest and fluttered them while she cried hoarsely. She seemed to resent the near approach of her mate or a helper to her nest during her absence; if one of them flew up with a billful of berries or other food while she perched nearby, she promptly returned to receive the offering while she covered her eggs. Rarely she accepted the meal among neighboring boughs. So well did her faithful provisioners keep her nourished that she spent nearly all of her recesses from incubation, which seldom exceeded ten minutes, preening her feathers and stretching her limbs on some convenient perch in sight of her nest. I doubt that she fed herself on her few brief excursions beyond my view.

Another beautiful jay is the Tufted (*Cyanocorax dickeyi*). Fifteen inches (38 centimeters) long, it is deep purplish blue, with white hindhead, nape, patch above the eye, cheeks, and all underparts posterior to the black chest, including the underside of the tail, the central feathers of which are broadly tipped with white. Its head is crowned with a high, spreading black crest, and its eyes are yellow. Adult males and females are alike, but yearlings differ in their lower crests, blue instead of white cheeks, and absence of the white spot above the eye. This jay, which was not discovered until 1934, inhabits a limited area in Mexico's Sierra Madre Occidental, between 4,500 and 7,000 feet (1,370 to 2,135 meters) above sea level, in southwestern Durango and adjoining areas of Sinaloa and Nayarit. In this rugged region of deep ravines filled with luxuriant subtropical vegetation, ridges wooded with pines and oaks, and towering cliffs, Tufted Jays were studied by Richard Crossin, chiefly during four months in 1964.

The Tufted Jays lived in flocks of four to sixteen individuals, including mature birds, two-year-olds, and yearlings. One flock of eleven split into groups of four and seven before nesting began. Flocks wandered widely over their broken terrain, finding much of their food among the abundant epiphytes, especially the bromeliads that they tore apart to reach the acorns, seeds, and small fruits that had lodged in their rosettes and the insects that hid in them. They plucked berries from thin twigs by hovering momentarily, trogonlike, before them. A variety of beetles and orthopterans filled out their diet.

Each breeding group contained a single mated pair, who dueted *sotto voce* and caressed each other's bills as no other group members did. In March the jays began to build their nests from 15 to 44 feet (4.5 to 13.5 meters) above the ground, amid dense foliage of trees in the wooded ravines. Nest construction was a group enterprise in which yearlings participated, although they and even two-year-olds assisted clumsily, dropping much of the material that they brought. Twigs for the foundation were broken from trees; brown or black rootlets for the lining were pulled from epiphytes; nothing, not even the many twigs that the birds had dropped beneath the nest, was gathered from the ground. After completion of the bulky, fairly substantial open bowl, freshly plucked green leaves were often arranged around its rim, as though for decoration.

The single breeding female of each group of Tufted Jays laid three or four eggs, which only she incubated. While sitting she was well fed by other flock members, who in the larger groups offered more than she could eat. When she did not respond to their arrival by rising slightly in the nest with quivering wings and begging calls, they gulped down what they had brought for her or carried it away. Her sessions of incubation, usually from half an hour to an hour, were short for a jay. Her absences, mostly lasting from twenty to twenty-five minutes, were probably not for foraging, since she was so well nourished by her attendants. During the second half of the eighteen-day incubation period, she spent from 80 to 95 percent of the day on her eggs. Her mate often rested for long intervals above her while she sat, but unlike the male Brown Jay, he did not often guard the nest in her absence. Even when not incubating, the female was usually watchful, and she came quickly to scold when Crossin tried to climb to the nest. As in other jays, the youngsters protested more vehemently than the older birds on these occasions.

Hatched naked, with tightly closed eyes and no trace of down on their dark pink, yellow-tinged skins, the young Tufted Jays developed slowly. About eight days after they hatched, their eyelids started to separate but were not fully open until the fourteenth day. Not until the eighteenth day did the nestlings' feathers cover them completely while they sat in the nest, although large areas of bare skin appeared when they were lifted from it. Soon after the nestlings hatched, all flock members brought them food, at first giving it to their brooding mother for delivery, later feeding them directly. At the age of about twenty-four days, the young jays left their nest and were led away by their group, with which, if they survive, most remain until at least the following year. The social system of Tufted Jays in Mexico closely resembles that of Brown Jays in Guatemala and Green Jays in Colombia.

John William Hardy, Ralph Raitt, and their teams of observers investigated the social systems of four related jays in Mexico and Nicaragua. All four species are black on the head, shoulders, and—more or less extensively—underparts, with different shades of blue on the back, rump, wings, and tail. Of these blue-and-black jays, the largest, the 16-inch (41 centimeter) Purplish-backed, or Beechey's, Jay (*Cyanocorax beecheii*), most resembles the foregoing species in its nesting habits. It is found only in the lowlands and foothills of the Pacific slope of northwestern Mexico, from southern Sonora to Nayarit. In this region where the dry season is long and severe, the jays live chiefly in the interior of thorny, deciduous woods, from which they emerge briefly in the early morning or late afternoon to forage in cultivated fields or mango orchards or to gather spilled grain along the railroad. For cooperatively breeding jays, the groups studied by Raitt and Hardy were small, consisting of only two to six individuals: a single mated pair accompanied by one or rarely two other adults at least three years old, sometimes one or two two-year-olds, and often one or two yearlings. Adults, with their yellow eyes and black bills, were readily distinguished from the yearlings with brown eyes and yellow bills.

In early May, while the dry season

still prevailed, the Purplish-backed Jays built, usually in one of the woodland trees that retained its foliage, nests that were disorderly piles of sticks lined with finer twigs. Most if not all group members brought materials and sat in the nest to shape it, but the breeding female did most of the work, especially in the final stages of construction. Then she laid from three to five eggs, usually the latter, which she alone incubated, fed at irregular and infrequent intervals by her mate. He also stood on the nest's rim or nearby to guard the eggs during about half of her absences, but he rarely waited for her return. All group members fed the nestlings, some yearlings bringing food as often as the father and more often than the mother. An attendant arriving at the nest with a meal often delivered a portion to another already there, then both fed the nestlings. Groups with helpers fed the brood between 4 and 5 times per hour, but an unaided pair brought food at the rate of only 3.5 times per hour. A group whose nest failed might soon try again, but no group had more than one successful nest in a year. Although Purplish-backed Jays are versatile foragers, seeking their food from the ground to the woodland canopy, in the woods and beyond it, the limited resources of an extremely arid environment apparently kept their groups smaller, and their reproduction more restrained, than those of closely related species.

Another blue-and-black jay, the 12-inch (30-centimeter) Yucatán (*C. yucatanicus*), inhabits deciduous forest and coastal scrub in the Mexican state of Tabasco, the Yucatán Peninsula, northern Guatemala, and Belize. Young in juvenal plumage are, for jays, amazingly different from adults, for their whole heads, necks, and underparts are plain white instead of black, and their bills are yellow rather than black, as in adults. In Campeche, Raitt and Hardy, with ten assistants, studied these jays during two breeding seasons and the intervening winter. As the nesting season approached, these jays were found in flocks of usually six to ten individuals, including adults, two-year-olds, and yearlings, who dwelt in harmony and peck-preened each other. Any individual of any age might give food to any other flock member. Often food was passed back and forth or through the bills of several birds before one of them swallowed it.

Unlike all the preceding species of jays, some groups of Yucatán Jays contained two or three breeding pairs with separate, more or less simultaneous nests. Each pair built a flimsy, shallowly cupped platform of sticks, lined with a few finer twigs or coarse vegetable fibers, usually in the lower part of the canopy of the small trees near the woodland's edge. Observations on building were limited, but at least three group members appeared to participate in it. Each female laid from four to six, most often five, eggs that were mottled pinkish buff, speckled with reddish buff, and very different from the eggs of other New World jays except the four blue-and-black species. It appeared that only a single female incubated, sitting for over 90 percent of the daytime, while she was fed by at least one or two group members, chiefly adults and only exceptionally yearlings, never two-year-olds. During her absences, often of less than three minutes and rarely longer than ten, another jay, probably her mate, stood silent and motionless on the nest's rim or nearby, guarding; but such vigilance was omitted at some nests.

When the nestlings hatched, after an incubation period of seventeen days, they were brooded by at least two black-billed adults. As the nestlings grew older and were brooded less, the number of their attendants increased. Since most of these jays were not banded or otherwise marked, the watchers could not learn just how many came to a nest. Most broods were fed by at least three adults, who brought nearly all their food, with yearlings and juveniles contributing a little. At the most active nest, the young were fed by no less than eight jays, including three adults, three yearlings, and two juveniles with white plumage. As at nests of other jays, one attendant often gave food to another, who in turn delivered it to the nestlings. On one occasion, it passed through three bills before a nestling received it! After the eggs hatched, some of the attendants spent much time inspecting the nest, especially the underside, and picking off small objects that were probably larvae of a fly that parasitizes nestlings, causing swellings beneath the skin.

An inhabitant of Central America from Guatemala to Nicaragua, the Bushy-crested Jay (*C. melanocyaneus*) occurs from sea level to 8,000 feet (2,440 meters) but is most abundant at middle altitudes. In the mountains north of Matagalpa, Nicaragua, Hardy found these jays abundant in coffee plantations where large trees of the cloud forest had been left for shade after thinning and removal of the smaller trees and the undergrowth. The birds also foraged in neglected fields and along cattle trails in the neighboring, relatively undisturbed forest, but they avoided closed forest. In June, while they fed fledglings from earlier nests, Bushy-crested Jays also attended nestlings or incubated three or four eggs in second-brood nests built of sticks and rootlets in the coffee shrubs. At each of two nests, incubation was shared by at least two adults, who while sitting were fed by no less than four other adults, with small contributions by two juveniles. With so much help, they kept the eggs almost constantly covered. At one nest, three older nestlings were fed by eleven adults and four or five juveniles, sometimes as often as twenty-eight times during an hour's watching.

In mixed palm and tropical deciduous forest in the Mexican state of Nayarit, Nelson San Blas Jays (*C. sanblasianus nelsoni*) lived in smaller groups (probably reduced by trappers) but had a similar social system, with two partly overlapping nesting periods in the same season. Here, also, two adults shared incubation. At one nest, the female who sat most of the time was fed by another who for brief intervals replaced her on the eggs. This second female was the principal caretaker of four fledglings in the vicinity, apparently her progeny. Two adult males also fed the incubating birds at this nest and likewise the fledglings, who also received meals from a yearling who rarely visited the nest.

The last blue-and-black jay to be investigated by Hardy and his associates was the southern race of the San Blas Jay (*C. sanblasianus sanblasianus*), which they studied intensively during five breeding seasons near Acapulco in the Mexican state of Guerrero. Here stable groups of thirteen to twenty-six birds one year or more of age, mostly at least three years old, were permanently resident in groves of coconut palms and neighboring patches of native woodland. Their ages were revealed by differences in plumage and the color of their eyes, which were brown until they were two years old, then changed through shades of green until they became pure yellow in three-year-olds and remained so for the rest of their lives. Individuals only a year or two old bred more commonly than in related jays, but most did not begin to nest until they were three years of age or older. Each group contained from six to ten breeding pairs that appeared to be permanently monogamous, plus a smaller number of nonbreeders.

Southern San Blas Jays built most of their nests in the crowns of the coconut palms, preferably those in the more open parts of the grove whose fronds did not interlock with those of neighboring palms, so that the nests were less accessible to flightless animals. The distance of each nest from its nearest neighbor in the same group ranged from 69 to 590 feet (21 to 180 meters), but a nest might be only 328 feet (100 meters) from a nest of another group. Each bowl of interlaced sticks was built chiefly by a female, assisted or at least accompanied by her mate, while other group members showed little interest and did not help. Only one female laid in a nest, and second broods were rare. During the seventeen or eighteen days that she incubated her three or four eggs, the female was fed only by her consort.

After the eggs hatched, helpers began to visit the nests, of which from five to ten might at one time be active in a single group. Except incubating or brooding females, most breeders assisted simultaneously at two or more nests in their group, and some brought food to every nest with young. While building their own nests, the jays fed the occupants of more advanced nests. In addition to feeding their incubating mates, males found time to help other pairs with nestlings. Breeding females served as helpers before they laid eggs or after their young fledged or their nests were lost. While feeding their own nestlings, however, the jays seldom visited other nests. Yearlings and other nonbreeders spread their attentions just as widely. These jays helped their parents when the latter had dependent young, but they began to assist at the first nest that produced nestlings, regardless of relationship. Some broods were nourished by as many as fourteen individuals, including the parents, neighboring breeders, and younger birds. Nests begun early in the season attracted more helpers than later ones, because the auxiliaries, whatever their status, preferred to feed the increasing number of fledglings rather than nestlings. Broods of older nestlings were fed at rates up to 4.5 times per hour or 1.8 times per nestling. The number of meals was not determined by the number of attendants so much as by the size of the brood and its need. The fewer the feeders, the harder each of them worked.

Two or three days before they fledged, the young San Blas Jays perched on limbs up to a few feet from their nest, then returned to it, often repeating this. When about eighteen to twenty days old and hardly able to fly, they fluttered to the ground, and for the next week they hid amid low vegetation near the nest tree or, running and fluttering, followed the calls of adults to better cover. A week after abandoning the nest, they could fly downward from low perches but not upward, and they soon tired. From this age onward they flew with increasing strength from tree to tree and remained above the ground. When members of Hardy's team carried fledglings, in full view of their attendants, into the home range of an adjoining group, the older birds refused to follow their calling young into foreign territory. However, the resident jays immediately adopted alien fledglings left among them, caring for them as though they were their own offspring.

San Blas Jays were careful not to cross the boundaries of their home ranges and avoided confrontations with neighboring groups. Their annual survival rate, as yearlings or older birds, was about 75 percent in most years. Most young and adult jays stayed in their natal group from year to year, and the few that emigrated were usually females. Even among cooperative breeders, these jays are outstanding in the number of individuals who interact in friendly, constructive ways.

Closely similar to the Southern San Blas Jay in its social system is the Gray-breasted Jay (*Aphelocoma ultramarina*), also known as the Mexican Jay, which from south central Mexico ranges northward over the highlands to central Arizona, southwestern New Mexico, and western Texas. About 12 inches (30 centimeters) long, this crestless jay has dull blue upperparts, wings, and tail and light grayish underparts, tinged with dull brown on the chest. As in a number of other cooperatively breeding jays, the bill, yellow or horn color in nestlings, darkens gradually until, at ages varying from slightly less than a year to more than two years, it becomes wholly black.

In the mountains of southern Arizona, where Gray-breasted Jays have been studied intensively for many years, chiefly by Jerram Brown and his associates, they frequent woods of evergreen oaks and pines at altitudes ranging from about 2,000 to 7,000 feet (610 to 2,135 meters) above sea level. Here they reside permanently in territorial groups of usually eight to twenty individuals, who make their presence known by loudly reiterated calls of *weet weet*. They appear not to perch or sleep in contact, but they forage or rest 2 inches (5 centimeters) or less from each other, which is closer than less social jays tolerate their companions. They eat insects, fruits, seeds, small vertebrates, and many acorns—holding them between both feet while with strong blows of their bills they peck them open to extract the embryos. They bury excess food in the ground and cover it with litter, or they hide it in crotches and crannies in trees.

Unaggressive, rarely engaging in boundary disputes with neighboring groups, Gray-breasted Jays lack a conspicuous dominance hierarchy; but an order of precedence could be demonstrated by providing a concentrated source of food, for example, placing it in a can set in the ground. In these circumstances, subordinates must wait until those higher in rank have eaten or have carried away all that their mouths and bills can hold, to be cached for future consumption. If present at the can when a higher-ranking jay arrived, the subordinate was unceremoniously displaced. However, at least in this artificial situation, the dominant individuals were not, as one might suppose from analogy with other cooperative breeders, the oldest jays but birds less than three years old. Apparently, priority was conceded to them because, less adept at foraging than more mature individuals, their need was greater. This is consistent with the behavior of other cooperative breeders who feed grown-up subordinates.

Gray-breasted Jays appear never to nest as isolated pairs without helpers. A single group may have on its territory up to six breeding pairs, with as many contemporaneous nests. Courting males feed females, but Gray-breasted Jays do not preen each other. In March or April they begin to build in trees, usually from 10 to 25 feet (3 to 8 meters) above the ground, rarely lower or higher. The jays strenuously break long twigs from oak trees that they interlace into a firm foundation for a cup of fine rootlets, which they line with soft dry grasses, horsehair, or cow hair. The nest is built chiefly by the female, with more or less help from her mate and rare contributions from other flock members, who crowd around, evincing much interest and at times sitting in the bowl to arrange materials.

The builders spend much time sitting, singly or the male and female together, in their completed but still eggless nest, to prevent the theft of its lining by other group members who are still building. The thieves, of both sexes, are so bold and aggressive that even the presence in the nest of both owners does not deter them; they reach under the sitting jays to pull out strands of lining. Although rarely a jay fights to protect its nest, usually it offers surprisingly little resistance. Sometimes an owner, sitting deeply in the bowl, begs from the robber with quivering wings, gaping mouth, and cries of *aaah*. The thief's response to this plea may be sharp pecks on the owner's head. Among colony-nesting birds that are not cooperative breeders, builders frequently pilfer materials from the unguarded nests of their neighbors; and in a garden where birds of several species nest close together, one often finds it easier to extract materials from the unwatched structure of a different species than to find what it needs at a greater distance; but we hardly expect members of a cooperatively breeding group to treat their associates so inconsiderately.

After the female has laid from three to five bright green eggs, exceptional in the jay family in their complete lack of markings, she alone incubates them, fed on the nest by her mate and other group members, sometimes as many as eight. One breeding female occasionally feeds another. When, after an incubation period of seventeen or eighteen days, the blind, downless nestlings hatch, more adults bring food at increased rates. At every nest watched by Brown and his coworkers, the young jays were fed by yearling and older helpers, regardless of their relatedness to their beneficiaries. Every member of a group of fifteen brought food to one nest, and at least twelve of these birds also attended a contemporaneous nest. Broods of four or five nestlings were fed at average rates of 4.1 to 5.3 times per hour. Although parents were engaged chiefly at their own nests, they found time to bring some food to neighboring nests, as they could well do, for their own nests received more meals from helpers than they themselves brought. While a mother brooded, her auxiliaries often presented their food to her for delivery to the nestlings. Among these helpers were aunts, uncles, nieces, nephews, siblings, and half-siblings of the young that they fed. Some served at nests of their grandfathers. Gray-breasted Jays appear never to feed nestlings or fledglings belonging to groups other than their own.

At the age of twenty-four or twenty-five days, when they can fly weakly from branch to branch, Gray-breasted Jays leave their nests. Those from different nests intermingle and are fed indiscriminately by all their parents and helpers. After another week, they are capable of strong, sustained flight. Thereafter, they tend to remain in their natal group for the rest of their lives, which may extend to eleven years or longer. Many do not breed until from two to five years old. Jays of both sexes often nest in their natal territory while their mothers and fathers and even their grandparents are still present. Thus, the more productive territories are inherited by direct descendants of both sexes from generation to generation. Jays who remain in their natal group nest with twice the success of those who migrate. However, emigrants rarely go farther than an adjoining territory. The sedentary habits of Gray-breasted Jays result in the helpers' close relatedness to the parents whom they aid and the young whom they attend. Nevertheless, there is sufficient movement between groups to prevent excessive inbreeding; many nesting jays are immigrants, and incestuous unions seem never to occur. Promiscuity has never been detected among these or, apparently, other jays.

Even more surprising than the theft of nest materials by birds so mutually helpful as Gray-breasted Jays is the occasional destruction of eggs by members of the same group. This has been called infanticide, which seems too strong a term, because an egg, although potentially a nestling, has a quite different appearance and lacks the begging nestling's strong appeal to parental feelings. Egg destruction by jays is less harsh than the infanticide practiced by such mammals as male lions and langurs, who kill the sucklings of newly acquired females so that they may the sooner become pregnant by the murderers of their babies. Jays who steal and eat eggs of members of their own group appear to be indulging without inhibitions in the practice, widespread in their family, of devouring eggs and young of other kinds of birds. Rare among Gray-breasted Jays, such misbehavior was detected at only 8 of 162 of their nests in a dozen years.

Most of the egg robbers were immigrants to the groups in which the incidents occurred, and they were members of groups with low reproductive success. Just as, at an earlier stage, the jays failed to resist with vigor those who extracted materials from their nests, so now both a guarding male and an incubating female were amazingly passive while others seized their eggs in their presence, merely calling, fluttering their wings, and gaping, never attacking the thieves. Rarely, Gray-breasted Jays mistreat nestlings of their own group. Two jays who pecked well-grown nestlings fed them a few days later, and these young fledged successfully. It appears that no society, of whatever animal, no matter how well integrated and pacific it may be, can wholly rid itself of miscreants; but its members may become too mild to resist them.

In the isolated race of Gray-chested Jays (*Aphelocoma ultramarina couchii*) in Brewster County, Texas, David Ligon and Sandra Husar watched groups of three birds feed nestlings, but this population has not been well studied.

All the foregoing species of cooperatively breeding jays occupy overlapping ranges from South America through Central America and Mexico (where most of them live) into adjoining regions of the United States. The one that we shall now consider, the eastern race of the Scrub Jay (*A. coerulescens*), is not only far separated from all these others but even a thousand miles distant from the rest of its species, which is widely distributed in western United States and Mexico and has only doubtfully been reported to breed cooperatively. Confined to limited areas of peninsular Florida, the eastern race of the Scrub Jay is becoming rare as its special habitat is converted into farms, grazing lands, and housing developments. A crestless jay about 11 inches (28 centimeters) long, its grayish blue crown, wings, and tail contrast with its grayish brown back. Its white throat is outlined by a necklace of blue streaks across its upper breast, and its more posterior underparts are whitish. Immatures are gray instead of blue.

In low, dense thickets of scrub oak, sand pine, and sand palmetto, interrupted by open, sandy areas where they forage on the ground, Scrub Jays are permanently resident in stable, territorial groups of up to eight or ten individuals. Although often difficult to watch amid the dense scrub, they frequent human habitations where they are fed, and they will become so tame that they eat from the hand. They harvest acorns and bury not only excess food but also small shiny objects such as spoons, keys, or thimbles. Although some of their calls are loud, harsh, or rasping, in repose they pour forth a medley of low, pleasant notes, the jays' nearest approach to song.

As in nearly all cooperative breeders, aggressive interactions between group members tend to be infrequent and mild; but by offering them peanuts from a mechanical feeder where only one could eat at a time, Glen Woolfenden, who with his colleagues has studied Florida Scrub Jays for many years, intensified competition among them and demonstrated a hierarchy. A group's breeding male takes precedence over all its other members. Male helpers dominate all females; but if they become too domineering toward the mate of the breeding male, he may intervene to protect her. The breeding female more mildly dominates female helpers. Juveniles in their first summer, subordinate to all other family members, squabble more than adults and seem to establish a hierarchy among themselves. As in Gray-breasted Jays, dominant individuals often permit young birds to take food that the former have found. In contrast to the usually placid coexistence of group members, boundary disputes between adjoining groups are frequent and noisy. In these confrontations, the alpha male takes the lead, and his consort, despite her inferior position in the hierarchy, displays greater zeal than her male helpers. On brief visits to bountiful sources of food beyond their territories, jays of different families eat pacifically only a few inches apart.

Each group contains a single, permanently monogamous breeding pair. In late February and March, when they prepare to breed, Scrub Jays choose sites for their nests only 20 inches to 10 feet (0.5 to 3 meters), rarely more, above the ground in a dense bush, often in or beside an open area amid the scrub, which offers few high sites. The open cup of coarse twigs, lined with tough fibers from the sand palmetto, is built by the male and female, unaided by other group members. After laying from three to five pale green eggs, spotted with rufous and cinnamon, the female incubates them alone, fed by her mate and rarely by other group members. Apparently, the reason why the helpers do not contribute more to the earlier stages of nesting is that they are rebuffed by the pair; but sometimes a persistent older helper, who has been long associated with the breeding male, is permitted to feed his mate. If a human approaches a nest while the female is absent but in view, she hurries to cover it, and she often sits so tightly that she can be lifted from it or posed by hand for a photograph. Such a visit to a nest often excites a more vigorous defense from helpers than from the parents themselves. Sometimes three or more members of a family pile into the nest, one on top of another, and assume defensive attitudes. After an incubation period of seventeen to nineteen days, the pink-skinned nestlings hatch blind and wholly naked.

With rare exceptions, Scrub Jays do not breed before they are two years old, and less than half do so at this age. While waiting to breed, the jays remain on their natal territory and help their parents. About half of the nesting pairs have auxiliaries, most often one, frequently two, rarely as many as five or six. Male helpers start to feed the nestlings a day or two after they hatch, females not until a few days later. The father and his older male assistants are the nestlings' chief provisioners, not only bringing food more frequently than the other attendants but carrying larger meals in their distended throat pouches. A three- to five-year-old male helper may outdo the parent in nourishing the young. Yearling males bring fewer and smaller meals than older ones, yearling females even less, and at a well-attended nest the mother brings least of all. Sometimes she leaves all the provisioning to her assistants. Nevertheless, at nests without helpers, the mother may bring nearly as much food as the father, or she may leave all the work to him while she spends most of her time on the nest. As a rule, an attendant of whatever category responds to the nestlings' begging and compensates for the deficiency of others, so that the young are not undernourished.

In most years, Scrub Jays raise no more than one brood, with the result that juveniles have no nest to attend. However, in an exceptionally favorable year, jays who have successfully reared one brood may nest again. In these circumstances, eight-week-old fledglings started to feed their younger siblings. After they began foraging for themselves, they often visited the nest and sat quietly while the father and older male helpers fed the nestlings. Then a female fledgling simulated feeding them. Two days later, she offered a small item to her mother; and after another day, her brother fed a nestling. These fledglings were still receiving from the older jays food that they sometimes delivered to the nestlings; but more often, after finding food for themselves or being fed, they flew to the nest and swallowed it there. Until the nestlings left their nest, these older siblings continued to bring them a little food.

Yearlings, who in the January following their birth can already forage as efficiently as adults, must also learn how to attend nestlings. At first, they fly to the nest and watch older birds feed them. After a few days of watching, they visit the nest alone and with empty bills go through the motions of delivering food to the brooding female. Soon they bring tiny portions, which as the days pass grow bigger. By the end of the nestlings' first week, these young jays, quick to learn, are contributing so much that the father, who may have no older auxiliaries, can relax his efforts to nourish them. With few exceptions, the helpers were offspring of the parents at whose nests they served.

In forty-seven reproductive efforts, each by one pair in one season, those jays without helpers produced, on the average, 1.1 fledglings and 0.5 young who survived to independence at the age of about three months. Fifty-nine similar efforts by pairs with helpers yielded an average of 2.1 fledglings and 1.3 independent juveniles. Thus, helpers more than doubled the production of juveniles per pair. This increased yield was not attributed to the better nourishment of nestlings with more attendants, for very few nestlings starved, and lighter fledglings survived about as well as those twice their weight. How, then, do helpers increase the productivity of nesting Florida Scrub Jays? Over 80 percent of all nest losses are caused by predators, including snakes, mammals, Fish Crows (*Corvus ossifragus*), and, rarely, the jays themselves. Parents and their helpers defend their nests by scolding, displaying, mobbing, and directly attacking predators. This defense apparently continues after the young leave their nests and scatter. The more numerous the attendants, the more constantly nest and young will be under surveillance by a jay quick to sound the alarm that will draw others, and the more effective their resistance should be. This, Woolfenden concluded after years of study, is the major way that helpers increase nesting success.

Yearling Scrub Jays enjoyed an annual survival rate of 88 percent. That of adults was only slightly lower, about 80 percent, which compares favorably with that of constantly resident small birds in the tropics. Individuals have been known to live for twelve years; one female continued to breed when at least eleven years old. Some Scrub Jays continue to help their parents for five years; none has remained for six years. Females, who do not inherit their natal territory, tend to venture forth at an earlier age and to go farther than males, whose chances of becoming established as breeders at home are better. When the urge to depart becomes strong, the young jay wanders through neighboring territories, seeking one that lacks a breeder of the same sex whom it can replace. If unsuccessful in its quest, the seeker returns home, as it may do after repeated fruitless excursions. A wandering jay who finds a vacant territory may join a partner from a different group and try to become established, but it may fail in the attempt and return to its birthplace—all very humanlike behavior. Scrub Jays rarely sever all contact with their natal group until they find another territory where they can nest. This possibility of rejoining their families and being accepted by them, instead of continuing to roam at random, is another of the multiple advantages of belonging to a cooperative group.

Male Scrub Jays often acquire a breeding territory by a process called budding. The territories of these jays expand as their families grow. When a territory becomes big enough, the oldest male helper may claim part of it for himself and attract a female from some other group; Scrub Jays do not mate with members of their own family. The section of the parental territory where the erstwhile helper rises to headship and breeding status might be regarded as his reward for helping enlarge the domain, not only by aiding in boundary disputes but also by greatly contributing to the nurture of young who augment the family's size and the area that it occupies.

The small (16-inch, 41-centimeter), all-black Northwestern Crow (*C. caurinus*) lives along the Pacific coast of North America from the Alaska Peninsula to northwestern Washington. Gregarious and almost fearless, it frequents the Indian villages and hardly moves away from children playing on the beach. As omnivorous as most crows, it scavenges with gulls along the shore and from a height, like the gulls, drops mussels upon the rocks to break open the shells. It flies inland to glean behind the plow and eats many berries in their season. Its nests of fine sticks and mud—lined with fibrous bark, dry grass, or deer hair—are built among coniferous trees near the shore, beneath boulders on the beach, or in holes in cliffs, sometimes in small, loose colonies. The four or five eggs are greenish, blotched with brown.

In five seasons on small islands off the coast of British Columbia, Nicolaas Verbeek and Robert Butler found a single yearling helper with 25 of 138 pairs of nesting Northwestern Crows. These helpers begged from the adult male and were fed by him at the rate of about four times a day. Sometimes an adult male eating a large item relinquished it to a hungry yearling. Although the female parent was frequently hostile to the helper, occasionally she preened the young bird, as did its father. Probably because of parental rebuffs, helpers were never seen to feed incubating females, and in only two of ten groups watched for many hours did they feed the nestlings or fledglings, twenty-three times in twenty hours at one nest, only once in the other group. The yearlings also stored food on the parental territories. Their chief contribution to the parents' nesting efforts appeared to be serving as sentinels at the nests and chasing intruders from the territories, alone or with one or both parents. On a territory they were dominant over all invading crows, who off the territory chased them. The yearlings also helped the parents mob or drive away such large predators as Bald Eagles (*Haliaeetus leucocephalus*), Peregrine Falcons (*Falco peregrinus*), Common Ravens (*Corvus corax*), and otters (*Lutra canadensis*).

Northwestern Crow nests with helpers yielded an average of 2 fledglings each, those without helpers only 1.2 fledglings. The yearlings' infrequent contributions of food to a few nests could hardly have increased the number of fledglings to this extent; but perhaps, as in the Scrub Jay, their assistance in guarding the nests promoted success. However, the most productive nests, which had nearly all the helpers, were situated near the beaches that were the major sources of the crows' food instead of farther inland, and this may have been responsible for both the greater productivity and the prevalence of helpers. Northwestern Crows have not progressed far in the development of cooperative breeding; unlike advanced cooperative breeders, after the nesting season they wander in large flocks instead of remaining permanently on their territories. Nevertheless, the presence of even rudimentary cooperation as far north as Canada is surprising.

On the islands where they watched yearling helpers, Verbeek and Butler found an unmated adult female Northwestern Crow with one blind eye and a deformed bill whose mandibles overlapped. In two consecutive years, this afflicted female associated with a mated pair and begged from the male, who in twenty-seven hours fed her twenty-four times and his nestlings sixty-six times, while his mate fed the young eighty-nine times but gave nothing to the dependent female. This observation strengthens the credibility of certain older reports of similar compassionate behavior. A Mr. Blyth told Charles Darwin that he saw Indian crows (probably the Jungle Crow, *Corvus macrorhynchus*) feeding her two or three blind companions, to which Darwin added that he had heard of an analogous case with the domestic cock. In *Mutual Aid*, P. Kropotkin wrote: "Brehm himself saw two crows feeding in a hollow tree a third crow which was wounded; its wound was several weeks old." Blue Jays (*Cyanocitta cristata*) are said by E. H. Forbush to care for the old and infirm, but details are not available. As reported in chapter 40, two widely separated observers saw wood-swallows feed disabled companions; and chapter 49 contains a record of a Black-headed Grosbeak (*Pheucticus melanocephalus*) with a deformed bill who was nourished by his mate.

The Pinyon Jay (*Gymnorhinus cyanocephalus*) is widely spread over western United States, where its preferred habitat is the pinyon pine–juniper belt of the mountains, although it also frequents woods of other pines and scrub oaks. Short-tailed, slender-billed, this 11-inch (28-centimeter) bird forages and roosts in large flocks. Its nesting colonies, sometimes spread over more than 100 acres (41 hectares), often contain scores of nests in coniferous trees or scrub oaks, occasionally two or three in the same tree but usually more widely dispersed. Each pair builds and attends its own nest, eggs, and nestlings until the latter are about two weeks old, after which neighboring jays are attracted by their loud, begging cries. Russel Balda and Gary Bateman saw as many as seven adults close to some nests, and a banded mother of four nestlings fed a neighboring brood. A male took food to two adjacent nests.

After leaving their nests, the young jays, attracted by their contemporaries' calls for food, hop or fly weakly into a loose crèche that usually contains from twenty to sixty fledglings spread over nearly 4 acres (1.6 hectares). Parents and their young recognize each other by voice, and each of the former feeds its own offspring, as in the crèches of penguins, flamingos, and other large birds. However, other adults are not wholly unresponsive to the fledglings' begging cries. An unbanded Pinyon Jay fed banded young from three different nests, and another gave food to fledglings of two broods. A female who had lost her own brood fed the offspring of more fortunate neighbors. Parents, chiefly males, give preferential treatment to their own progeny, feeding them first and more frequently than young of other parents who may be calling nearby. They also preen these fosterlings. Crèches are guarded by sentries, not only parents but also nonbreeding yearlings, whose cries of alarm draw other adults to help them mob and harass approaching predators.

As in Jungle Babblers (*Turdoides striatus*), young Pinyon Jays in crèches are more quarrelsome than adults; but after they join the fall and winter flocks, the aggressiveness of these youngsters declines dramatically. Throughout their first year, the young jays appeared to enjoy a special status in the flock; at a feeder they did not behave like subordinates but were indulged by most older birds. Among Gray-breasted Jays and Florida Scrub Jays we noticed this same deference to the young, which appears to compensate for their inferior competence in foraging and may increase their survival. The gray yearling Pinyon Jays were not seen to feed nestlings or fledglings, but one of them gave food to an incubating female.

Widely distributed in open country across Africa south of the Sahara and north of the equator, the Piapiac (*Ptilostomus afer*) is a slender, long-tailed, black bird with a short, thick bill and beautiful purple or violet eyes. Juveniles, only slightly duller than adults, have brown eyes and pink bills with dark tips, which do not turn black until they are about a year old—a condition not unlike that in many New World jays and probably of similar significance. In flocks of ten or more individuals, Piapiacs forage chiefly while they walk or run swiftly over the ground. Often they accompany cattle, goats, and sheep, perching upon their backs and catching insects that they stir up from the grass. They pluck vermin from the skin of elephants. In addition to insects and other invertebrates, these black birds eat the pericarp of oil palm fruits. Their voices are shrill and squeaky. Adults preen each other and immature birds. Hardly afraid of humans, Piapiacs rest on dwellings and roost in palms or other trees in or near villages.

In branching trees or palms, Piapiacs build deep cups of twigs, long strips of palm fibers, and grass stems, lined with finer palm fibers. Nests have been found with three to seven pale blue or greenish blue eggs, spotted and blotched with shades of brown and pale lilac. The larger sets may be laid by two or more females. Llewellyn Grimes watched one immature and four adult Piapiacs feed three young so recently out of the nest that they could not fly. In an hour, five adults and two or three immature birds visited the nest site. Elsewhere, more than two birds defended a nest with young. This corvid of doubtful relationship appears to be a regular cooperative breeder that invites careful study.

Another African bird that possibly nests cooperatively is the Zavattariornis (*Zavattariornis stresemanni*), a corvid with a pale bluish gray body, whitish head, and black wings and tail that inhabits a small area of parklike thorn-acacia country in southern Ethiopia. Three adults have repeatedly been seen to emerge through the vertical spout that forms the entrance to the roughly globular nest of thorny twigs, about 2 feet (61 centimeters) in external diameter, where up to six eggs are laid.

The Siberian Jay (*Perisoreus infaustus*), a grayish, dark-headed bird with bright reddish wing patches and outer tail feathers, ranges widely through the boreal forests of Eurasia from Norway to far eastern Siberia and southward to the Altai Mountains. A third individual is often tolerated in a breeding territory and may replace a member of a breeding pair who dies. At a nest watched by F. Lindgren, a female and one of these extra birds sat together on the eggs throughout the incubation period. Even birds that regularly share a nest, such as anis, never occupy it together except very briefly, but among far northern birds such simultaneous incubation by two individuals, apparently the breeding female and a helper, is probably not as unusual as it seems.

On snowy days in Canada, Louise Lawrence watched two Gray Jays (*P. canadensis*) sitting together in a nest, one above the other, and taking turns at being the bottom bird in contact with the eggs. In the freezing weather when these boreal jays often nest, two simultaneously sitting birds should help keep the eggs—and each other—warm.

In captivity, a young Jackdaw (*Corvus monedula*) fed younger individuals. While the city of Washington was covered with snow and ice, a Common Raven in the National Zoological Park passed food through the bars of its outdoor cage to a free Black Vulture (*Coragyps atratus*), the only record of interspecific helping that I have found in this family.

REFERENCES: Alvarez 1976; Balda and Balda 1978; Balda and Bateman 1971, 1973; Bent 1946; J. L. Brown 1963, 1972; J. L. Brown and E. R. Brown 1980, 1981, 1984; J. L. Craig et al. 1982; Crossin 1967; Darwin 1871; M. Davis 1952; Forbush 1927; Goodwin 1976; L. G. Grimes 1976; S. A. Grimes 1940; Gross 1949; Hardy 1961, 1976; Hardy et al. 1981; Kropotkin 1902; Lawrence 1947; Lawton and Guindon 1981; Ligon and Husar 1974; Lindgren 1975; Moore 1938; Morrison and Slack 1977; Nice 1943; Raitt and Hardy 1976, 1979; Raitt et al. 1984; Skutch 1935, 1953a, 1960; Trail et al. 1981; Verbeek and Butler 1981; Woolfenden 1974, 1975, 1978; Woolfenden and Fitzpatrick 1977, 1978.

26.

MUDNEST BUILDERS

Grallinidae

Mudnest builders are a family of only three genera with four species, which range in length from 8 to 18 inches (20 to 46 centimeters) and lack bright colors. The Magpie-Lark (*Grallina cyanoleuca*), boldly patterned in black and white, is a familiar bird over much of Australia, where it finds most of its food on the ground, chiefly at the brink of ponds and streams. A related species, the New Guinea Mudlark (*G. bruijni*), dwells along torrents in the forested mountains of that island. The other two species, confined to Australia, are the subjects of this chapter. The family is named for the solid, bowllike nests of mud strengthened with grass or hair that all its members build on stout limbs of trees.

The black plumage of the biggest member of the family, the White-winged Chough (*Corcorax melanorhamphus*), is relieved by a large white patch on each wing, conspicuous when they are spread, and by the bright red eye of fully adult birds. Its legs and strong, decurved bill are also black. In open woodlands and fields of eastern and southeastern Australia, choughs live in groups of two to twenty individuals, most often about seven, who walk over the ground in extended formation, covering 3 or 4 miles (5 or 6 kilometers) daily, raking up the litter and turning over branches to disclose the insects and other small creatures lurking beneath them. If one finds an especially productive spot, all converge upon it and share the bounty without quarreling. When threatened by an aggressive Black-backed Magpie (*Gymnorhina tibicen*), the scattered foraging choughs quickly cluster into "a screaming piebald mass of black and white with a dozen or more crimson eyes, gaping bills and flashing white wing-patches." By such prompt mutual defense, the weakly flying choughs hold the more agile enemy aloof.

In the breeding season, each group of choughs defends a territory of about 50 acres (20 hectares), but in winter, when they include more grains and fruits in their diet, they wander more widely over a home range of 3 or 4 square miles (800 to 1,000 hectares), often mingling with other groups in favorable areas. If such a large aggregation of foraging choughs is disturbed, the birds separate into their family groups and disperse in different directions. Weak-winged birds, they fly reluctantly, preferring to ascend to a nest high in a tree by hopping or flitting from bough to bough rather than by direct flight.

White-winged Choughs are long-lived birds who usually do not breed before their fifth year, when their brown fledgling eyes have at last become wholly red, but physiologically they may be ready at an earlier age. Males and females occur in the population in approximately equal numbers. Typically, a group consists of one or two adult males and females, with a variable number of younger birds of both sexes. These groups grow chiefly by their own reproduction and tend to be stable over the years, but they usually disintegrate if they lose their adult male, the linchpin of the family. Dispersed individuals are gathered into new groups by adult males. Females who migrate from one group to another are often accompanied by immature birds of either sex, who may be their own progeny, with the result that even newly formed groups usually consist of four to six birds; breeding groups of only pairs and trios are rare.

Members of a group dwell in great harmony. Ian Rowley, who for ten years studied White-winged Choughs in farmland around his home near Canberra, characterized them as "mild, slow-maturing, slow-breeding, weak-flying birds, ill-adapted to cope with new predators such as the fox and feral cat." In a decade, he saw only twenty-five instances of aggression between groups and three instances within groups. Dominance, rivalry, and disputes over food were not evident. At night, the choughs generally roost well above the ground within the dense foliage of a large eucalyptus tree, all members of the group perching side by side in close contact, usually facing into the prevailing wind.

The absence of an especially close association between any two group members, and of jealousy or rivalry, makes it difficult to determine the choughs' mating system, which might be polygyny, polyandry, or promiscuity. However, the occasional laying of two eggs in a nest on the same day and the presence of exceptionally large clutches and of eggs of strikingly different appearance in a set are clear indications that two or more females sometimes deposit their eggs together. For the support of their open, bowllike nest of hardened mud, bound together with vegetable fibers, the choughs choose a stout horizontal branch about 33 feet (10 meters) above the ground. Most often an old nest is refurbished. In the early morning, with great social excitement including much calling and displaying, all members carry material to the nest, deposit it there, and shape the structure. Sometimes they queue up while waiting for their turns at the nest. Birds less than one year old build inexpertly and occasionally damage a soft, unfinished nest by their clumsy movements. When two years old, the choughs build as diligently and skillfully as their elders. Since each new layer of mud must harden before more is added, building proceeds slowly; it seldom takes less than a week and may be prolonged for several months, particularly if a dry period intervenes.

A White-winged Chough generally lays from three to five eggs; the few sets of six to nine are probably the product of two or more females. During an incubation period of 19.3 days, all group members take turns on the nest, yearlings infrequently. Their sessions on the eggs range from 2 to 244 minutes and average 50 minutes. Often the incubating bird relinquishes its post quickly and quietly to another coming for its turn, but sometimes it is reluctant to depart. When this occurs, the newcomer displays to it, fanning out and wagging its tail, spreading and waving its wings to reveal the large white patch on each, while its eyes bulge out as two scarlet orbs and it utters piping notes. Before long, the bird on the eggs is impelled to rise and respond with a similar display, whereupon the newcomer pokes its head beneath the sitting bird and replaces it on the eggs.

After the eggs hatch, group members of all ages brood the young and bring them worms, caterpillars, centipedes, moths, frogs, and lizards, all collected on the ground and carried, unless very large, several together, held crosswise in the bill. The rate of feeding is determined not by the number of feeders so much as by the nestlings' needs. Since larger groups do not provision the nestlings at a more rapid rate than smaller groups do, their members do not work so hard. Until the nestlings are feathered, they are brooded almost continuously, and thereafter the attendants take turns guarding them from a nearby perch—being quick to sound the alarm call if danger threatens.

When, at an average age of 27.6 days, the young choughs leave the bowl of mud, they cannot fly properly and appear to need encouragement before they flutter down with frantically beating wings and land heavily. For the next week or more, they remain flightless but run strongly and can climb into bushes or up sloping trunks. Those that leave the nest first cannot return to be fed and to sleep with their siblings, who may remain for two or three days longer. In this interval, group members divide the responsibility: some continue to feed the remaining nestlings, while others take care of the fledglings and may not come near the nest. After all the brood has fluttered to the ground, the group is reunited, and all its members continue to feed the young indiscriminately for at least a month. When, as rarely happens, a large group undertakes to raise a second brood, some members take charge of the young of the first brood while others give their attention to the new nest.

Despite the presence of helpers, White-winged Choughs nest with poor success. Of 298 eggs laid in 74 nests, only 143 (48 percent) hatched. More than two-thirds of the nestlings died in the nest, mainly because the smaller members of a brood were inadequately nourished. On the average, each nest yielded only 1.14 fledglings. Unlike many other cooperative breeders, such as anis and woodpeckers, choughs cannot dispense with their helpers; they are absolutely necessary because pairs and trios produce negligible numbers of offspring. Four-fifths of the choughs in Rowley's study lived in groups of four to eight individuals, which is the optimal size for reproduction. Groups of four or five rear to fledging 50 percent more nestlings for each attendant than do groups of six to eight, indicating that the larger groups have a superfluity of helpers attending nestlings. However, after

Apostlebird

leaving the nest, young choughs survive to adulthood at four years or more very much better in groups of six to eight than in groups of four and five. Like many other birds, they suffer the heaviest mortality in their first year of life.

Although the adaptable black anis, in some ways the choughs' counterpart in the New World, have profited by the replacement of forests by agriculture and pastures with grazing cattle, the White-winged Choughs of Australia have been adversely affected by the changes wrought by European colonization, and Rowley fears that they are on the road to extinction—which would be a great pity. However, their ability to adapt to the new plantations of introduced pine trees and seek their food among fallen needles may be their salvation.

Rather similar to the White-winged Chough in its social habits is the Apostlebird (*Struthidea cinerea*), a 13-inch (33-centimeter) gray-brown bird with a stubby bill. It is named for the groups of about twelve individuals, the number of apostles in the New Testament, that live in open country west of the Great Dividing Range, from northern Queensland to northwestern Victoria. Walking over the ground with tails swinging from side to side, running briefly, occasionally hopping twice their own length, loudly chattering groups of Apostlebirds gather insects, worms, seeds, and leaves of grass, clover, and other green herbage. With their bills they turn over or push aside the ground litter. They capture flying insects by jumping into the air or hawking them from treetops. Leaping and scrambling up trees, they probe the bark for small invertebrates. Far from shy, they follow close to human feet, take grain or soaked bread from the hand, and when hungry solicit food by pecking the arm or empty hand of a motionless person. But they are wary and will not permit themselves to be touched.

Group members crowd together on a horizontal branch, or more rarely on the ground, and preen each other, anointing their feathers with oil nibbled from their own or another bird's gland at the base of the tail. Fledglings press close to adults and are frequently preened by them when they lift their wings or present their napes, at the same time grooming the plumage of their own breasts and abdomens. Or a male nestles between two females, who alternately preen his nape or throat, sometimes continuing for a quarter of an hour.

Like babblers, ground hornbills, and other highly social birds, Apostlebirds play, especially when cooled in summer by a drop in temperature or a shower or on a warmer day in winter. They hop quickly up a ladder only to glide down, then repeat the performance again and again. Or they follow-the-leader around a tree, each crouching low and trying to nip the tail of the bird ahead. They dart away, then return with loud calls. In a rougher game, they strike out with their feet and try to turn an opponent on its back while they exchange sharp pecks. Or the dominant male may voluntarily lie on his back in a relaxed attitude while other birds peck his abdomen for a while. Then he jumps away from them and leads another chase around a tree. In what is perhaps redirected aggression, the Apostlebirds pull twigs to the ground and hold them with their feet while with their bills they tug and twist until the branches break. They strip off leaves and tear apart flowers. A young male may challenge a dominant male by approaching him with a leaf held high in his bill, resulting in a series of tugs-of-war, in which the leaf is torn between them into successively smaller fragments.

A breeding group of Apostlebirds consists of three to nineteen individuals, usually about nine, including a dominant male with often two or three adult females and young, who remain for several years with their parents on a defended territory. As the dominant male flies, loudly calling *kree*, to confront a trespassing group, he is followed by the members of his clan, all shouting *chee-ow*. With fluffed-out feathers and spread tails, the defenders pick up and carry in their bills grass, earth, or pebbles, much as Green Wood-hoopoes (*Phoeniculus purpureus*) wave "flags" in territorial encounters. The intruders usually withdraw without a fight.

Together the members of a group build a bowl of hardened mud, reinforced and lined with grass, most often high in a tree. Here from four to eight pale bluish eggs, with dark brown and purplish markings, are laid, the larger sets by at least two females. Juveniles of the season's first brood help build the second nest, then join other group members in attending the young. Unfortunately, no detailed study of the Apostlebirds' nesting, comparable to Rowley's study of the White-winged Choughs, is available.

Merle Baldwin, who for over a decade watched Apostlebirds around her home near Inverell in New South Wales, found them intimately associated with White-winged Choughs. While the choughs forage over the ground, digging deeply for large grubs with their long bills, the short-billed Apostlebirds often follow them, gleaning smaller items from the loosened soil. Choughs building their first nests in July appear almost to be supervised by Apostlebirds, who do not breed until October. They sit beside or in the choughs' nest arranging materials, then watch close by while the larger birds incubate. When rain falls into the bowllike nests, either species may help the other drain out the water; but they do not cooperate with Magpie-Larks in this manner. Apostlebirds warn choughs of danger and join them in attacking predators; the choughs' cries of alarm bring Apostlebirds to aid them.

In contrast to the other two genera of mudnest builders, Magpie-Larks nest in permanently mated, territory-holding, monogamous pairs, apparently always without helpers. Both parents build, incubate, and rear the young.

REFERENCES: Baldwin 1974; Rowley 1975, 1976, 1978.

27.

AUSTRALIAN BUTCHER-BIRDS AND BELL-MAGPIES

Cracticidae

The ten species of this small family are confined to Australia, New Guinea, and nearby islands, where they inhabit open woodlands, brushy places, pastures with scattered trees, farms, city parks, and suburbs. From 10 to 20 inches (25 to 51 centimeters) long, they are mostly black, white, and gray, often strikingly pied. Stoutly built, they have large, strong bills with prominently hooked tips. Like the northern birds of a different family for which they are named, the Australian butcher-birds catch insects, lizards, mice, and small birds, which they impale upon thorns or ram into crevices in trees while they tear the victims into swallowable fragments. Members of this family are highly esteemed songsters who join in delightful choruses. In trees, often high above the ground, they build open nests of sticks lined with softer materials and lay from two to five eggs, highly variable in color but usually spotted or streaked with darker shades.

The Gray Butcher-bird (*Cracticus torquatus*) is found over most of Australia and Tasmania, where it lives in pairs that defend their territories throughout the year and sing together. The incubation period of the three or four pale brown eggs, marked with reddish brown, is twenty-three days, and the young remain in the nest for twenty-five or twenty-six days. In New South Wales, Ian Rowley watched two adults and a

Western Magpie

juvenile feed nestlings. For the boldly black-and-white Pied Butcher-bird (*C. nigrogularis*), widespread in inland Australia, we have two reports of nests attended by trios, consisting of two adults who were doubtless the parents plus a juvenile.

The large bell-magpies, who resemble the true magpies (*Pica*) in their black-and-white plumage, carol beautifully with bell-like voices. They are highly sociable birds that forage mainly on the ground, where they probe for insects with long, sharp bills and overturn cow pats and other small objects to see what lurks beneath. The Western Magpie (*Gymnorhina dorsalis*) of southwestern Australia lives in territorial groups of three to twenty-four individuals. A single group may have several females simultaneously attending large, bowl-shaped nests that are occasionally built of several pounds of wire, including hundreds of pieces from a few inches to 4 feet long! When provoked in the breeding season, magpies often attack humans, including innocent passersby, so vigorously, inflicting such painful scalp wounds, that in several Australian cities they have been shot to protect the public. Males assist little at the nests, but after the young fledge, adult males and immature birds may help feed them for up to six months. Similarly, in the Black-backed Magpie (*G. tibicen*) of eastern Australia, the female builds the nest, incubates, and supplies most of the nestlings' food, but all group members help to feed fledged young as well as to defend the territory. Despite the complex social systems of Australian magpies, cooperation, other than in territorial defense, is poorly developed in them.

References: Rowley 1975, 1976.

28.

TITMICE

Paridae

Most of the forty-five species of titmice inhabit the Old World, in Eurasia, Africa, and Indonesia. In the western hemisphere they are found from Alaska to the highlands of Mexico; none reaches Central or South America. They are small birds, from 3.5 to, rarely, 8 inches (9 to 20 centimeters) long, with soft, thick plumage boldly patterned in black, white, gray, brown, yellow, and blue or else plain gray. Some titmice wear upright crests. In lisping flocks, they roam through woodlands, busily climbing and hopping over trunks, branches, and twigs in search of small insects and spiders. Nuts and berries fill out their diet. Their songs at best are bright and cheerful rather than brilliant and complex. In old woodpecker or barbet holes, cavities resulting from decay or sometimes carved by themselves in soft wood, as well as birdhouses and diverse crannies, they build bulky nests of soft materials and lay from four to a dozen, rarely more, white or tinted and often spotted eggs, which the female incubates while her mate brings her food. Both parents feed the nestlings.

The only member of this family known to breed cooperatively is the Black Tit (*Parus niger*), a widespread, common resident of moister savanna woodlands of southern Africa. A white wing bar contrasts with its dark plumage; females' underparts are paler than those of males. For twenty-seven months Warwick Tarboton studied a color-banded population of these titmice in central Transvaal. Of nineteen breeding units, eight consisted of simple pairs, nine of two males and a female, one of three males and a female, and one of four males and a female, all adult birds, among which males outnumbered females in the ratio of 1.7 to 1. In each group, the single dominant breeding male was paired with the only female; other males, when present, were nonbreeding helpers. These groups were permanently resident on all-purpose territories, the area of which increased with the number of occupants from 62 to 74 acres (25 to 30 hectares) for simple pairs to 89 to 118 acres (36 to 48 hectares) for three or four adults. While foraging over these large territories, groups traveled at an average rate of 6 miles (10 kilometers) per day, often meeting neighboring groups at the boundaries. The disputes that then arose involved much calling, chasing, supplanting, and physical fighting between members of rival groups. Black Tits' confrontations with their neighbors appear to be less ritualized, more violent, than such encounters commonly are among cooperative breeders.

Black Tits nested from October to December, the spring and midsummer of the southern hemisphere. As the date of egg laying approached, the alpha male frequently fed his mate, who received the food with quivering wings and a rasping, begging call. She also solicited from the helpers, who tried to respond with gifts of food but were repeatedly chased away by her jealous consort. However, when two helpers were present, one sometimes managed to feed her while the head male pursued the other.

For their nests, the tits chose natural cavities from 6 to 16 inches (15 to 41 centimeters) deep, usually 6 to 10 feet (2 to 3 meters) above the ground in the main trunks of the small trees of the savanna. The female alone built the nest, accompanied by her mate on the ground while she collected vegetable fibers and mammalian hair to form a thick pad on the bottom of the hole. Here she laid four or five eggs, which she incubated without help, sitting from about half an hour to an hour at a stretch and keeping her eggs covered for 76 percent of the daytime. After the eggs were laid, the dominant male, now assured of his paternity, freely permitted the helpers to feed his mate. At intervals he or they called her from the nest to receive a meal, after which she accompanied them for outings that continued for about half an hour, while the nest remained unattended. On returning, especially early in the incubation period, she sometimes brought more material for her nest. If disturbed, she sat tightly on her eggs or nestlings and tried to intimidate the intruder by suddenly recoiling her head and emitting a loud hiss—a snakelike display employed by other species of titmice and hole-nesting birds of other families.

As she had incubated alone, so the female brooded her nestlings and delivered to them food brought to her by her mate and helpers. As the nestlings grew older, she left them alone for increasingly long intervals while the attendants fed them directly. At four nests, the father was the chief provider. At a

Black Tit

nest without auxiliaries, he brought two-thirds of the meals and the mother brought one-third. At a nest with two helpers, the father brought 38 percent of the food, the mother 15 percent, and the helpers respectively 37 and 10 percent or nearly half of the nestlings' meals. The overall rate of feeding averaged 8.9 meals per hour, with no correlation between the feeding rate and the number of feeders or the size of the brood. The unaided parents of a brood of five were stimulated by their nestlings' continual begging to bring food more frequently than any individual member of a larger group did; but the cries of these nestlings, audible at least 50 yards (46 meters) away, attracted a predator who ate them. Nestlings with more attendants did not need to make themselves conspicuous by calling for food.

After leaving the nest cavity, loudly begging fledglings continued to be nourished by all members of their group. In their second week in the open, the young Black Tits began to find food for themselves. By their seventh week, they ceased to solicit food, lost the yellow corners of their mouths, and were no longer fed. Tits still dependent on the adults moved rather frequently from one group to another, a transfer that apparently occurred during the disputes between families at their common boundaries. In these altercations, adults often pursued each other for several hundred yards, leaving the young behind. When the engagement was over, a fledgling often found itself in the territory of the opponents, solicited from them, and was promptly adopted. A consequence of these interchanges might be that in the next breeding season a yearling male helped his foster parents rear nestlings not his siblings and possibly only distantly related to him—a fact that must not be neglected when theorizing about cooperative breeding.

Because their parents raised only one brood in a season, juvenile Black Tits lacked opportunities to serve as helpers. When about four months old, the tits molted their body plumage and became indistinguishable from adults. Now the young males helped defend their group's territory and were repelled from the territories of neighboring groups. Two subordinate males, already adult when banded, helped in the same group in two consecutive breeding seasons. Females, who could breed when one year old, were less settled than males. They remained in their natal groups for up to nine months, then dispersed. Even an adult female might leave the territory where she had nested, to become the breeder of an adjacent group. One female abandoned her territory for about six weeks, probably searching for a vacancy that she did not find, then returned to the territory, where she had not been replaced, and nested there in the following breeding season.

Unassisted pairs of Black Tits reared an average of 0.88 young per season; pairs with helpers reared 1.55. The productivity of nests did not correlate with the number of attendants. It is instructive to compare the reproductive rate of Black Tits with that of the Great Tit (*Parus major*) of Eurasian broad-leaved woodlands. The latter, breeding in unassisted pairs, lay sets of nine or ten eggs instead of the Black Tits' four or five and rear an average of 2.8 young per adult per year, six times the Black Tits' yield of 0.47 young per adult per year. The northern bird must rear many more fledglings to compensate for its far greater annual mortality.

Reports from widely separated localities suggest that the gray, tawny-flanked, high-crested Tufted Titmouse (*P. bicolor*) of the woodlands of eastern and central United States has helpers not infrequently, but hardly enough to be included among cooperative breeders. In Maryland, a banded, unmated yearling helped for two weeks to feed the three nestlings of its mother, whose mate had died and who was now paired with a different male. Although the young bird's attendance was less regular than that of the nestlings' parents, it brought food at least 89 times, while the father delivered at least 120 meals and the mother 126. The latter was friendly with the helper, but the male tried to drive it away. In Oklahoma, a second female, of unknown relationship to the breeding pair, assisted in feeding their four young both in the nest and after they fledged. In Tennessee, an unmated yearling of undetermined sex helped a pair, apparently its parents, to nourish their fledglings. In the same state, parents rearing a second brood were assisted in feeding the nestlings by two other titmice who appeared to be young of their first brood. Another pair of Tufted Titmice was helped at the nest by a Carolina Chickadee (*P. carolinensis*).

In California, a pair of Williamson's Sapsuckers (*Sphyrapicus thyroideus*) and a pair of Mountain Chickadees (*Parus gambeli*) nested in holes in the trunk of a living tree with a decayed heart, the chickadees in the higher cavity. At some time during the nesting, the thin partition between the two holes collapsed, dropping the chickadees' nest into the sapsuckers' chamber. When this nest was found, the sapsuckers were carrying ants to their nestlings and the chickadees were bringing insect larvae to the young sapsuckers, the contents of their own nest having vanished. Like other nestling woodpeckers, young sapsuckers grab food abruptly from their parents' bills instead of presenting gaping mouths and waiting for food to be placed in them, as is usual in passerines. Their rough behavior troubled the helpful chickadees, who sometimes retreated from the doorway, only to return again and again to deliver their offerings to the adopted brood. This little drama was abruptly terminated by collecting some of the actors.

In England, a female Great Tit whose mate had died led her eight fledglings into Len Howard's garden, from which she had earlier been driven by the pugnacious resident male. Now from time to time he gave to the widow's young caterpillars that he had gathered for his own nestlings. Later, a pair of Great Tits adopted eight young fledglings, grandchildren of the helpful male. Two juvenile male Great Tits from the same brood were inseparable playful companions. After one of them broke a leg and their games ceased, the sound tit placed food within reach of his injured brother.

Also in England, a pair of Blue Tits (*P. caeruleus*) built a nest in a box, on top of which a pair of Robins (*Erithacus rubecula*) already had a nest. The female Robin laid five eggs and the tit laid three. After the Robin's eggs hatched, the tits covered their own eggs with feathers and fed the nestling Robins. At first the Robins and their helpers fought a little, but soon the two pairs settled down to attend the nestlings in concord. After the Robins fledged, the Blue Tits laid another set of seven eggs over the original three and raised a brood.

References: Brackbill 1958; M. F. Davis 1978; Howard 1952; Laskey 1957; Lonsdale 1935; Pullman 1970; Russell 1947; Tarbell 1983; Tarboton 1981; Williams 1942.

29.

LONG-TAILED TITS AND BUSHTITS

Aegithalidae

Six of the seven species in this small family range through the Old World from the British Isles to Japan and Java; the seventh is confined to North America. Only 3 to 6 inches (8 to 15 centimeters) in length, these long-tailed, short-billed birds are black, white, gray, rufous, or pinkish. In loquacious flocks they move through woodland and thicket, gathering small insects and spiders from foliage and twigs while they cling in every imaginable position. In elegant globular or pouchlike nests with a side entrance, they lay from four to twelve plain white or spotted eggs that both sexes incubate. Both feed the young.

The Bushtit (*Psaltriparus minimus*) is a tiny grayish bird about 4 inches (10 centimeters) long. The black-eared color phase, which occupies the southern part of the species' range, from southwestern New Mexico and western Texas over the highlands of Mexico to Guatemala, was long classified as a different species (*P. melanotis*). The black-faced male has dark eyes; the female, who has only a small black patch behind each ear, has pale yellow eyes. Both sexes have black bills and legs.

In Guatemala I found Bushtits from 5,000 to 9,000 feet (1,525 to 2,745 meters) above sea level. In January they roamed through light woods of oaks, pines, and alders and over bushy fields and pastures in flocks of a dozen to a score of individuals. Most of the time they remained well hidden amid foliage, from which they emerged only long enough to pluck a small creature from the exposed tip of a branch. When, occasionally, they became more visible, I counted four to six black-faced, dark-eyed individuals for every gray-cheeked, yellow-eyed female. Once, in a garden, I watched a flock of eleven black-cheeked Bushtits with only one gray-cheeked individual.

Without any song to proclaim territory, pairs separated from the winter flocks and spread over the mountains to breed. On a bushy slope high on the Sierra de Tecpán, in early March, I found the most charming nest that I had ever seen. The delicate, pear-shaped pouch, 6.5 inches (16.5 centimeters) long, hung 8 feet (2.4 meters) up from the fork of a thorny, downy-leaved shrub of the potato family (*Solanum mitlense*), amid its big, pale lavender blossoms. The whole surface of the pouch was covered with finely branched, gray foliaceous lichens, bound together with cobwebs. Mixed with the lichens were tiny pellets of down and bits of the egg cases of spiders. At the top of the nest, between the arms of the supporting fork, a round, sideward-facing doorway opened beneath a protecting hood.

The nest's fabric was still thin and delicate, and the Bushtits were lining it. As at another nest that I watched, both sexes participated about equally in this task, often arriving together with down in their bills, which they always carried inside. Sometimes, while attaching their materials, they made the thin wall shake until I feared it would rupture, but my misgivings were groundless. On the bottom they deposited much vegetable down to form a thick cushion, above which they laid a coverlet of soft feathers. After I found the exterior already finished, the pair took nearly three weeks just to line the pouch. The whole work of construction could have occupied no less than a month. As soon as the nest was ready and before an egg was laid, the industrious pair slept in it on chilly mountain nights.

Each of the three nests that I found received four pure white eggs, laid on consecutive days. Alternating in the nest, the male and female kept the eggs rather constantly covered in the cool early morning but were neglectful during the middle of sunny days, when doubtless they found their snug pouch uncomfortably warm. Nevertheless, they continued to bring more down to it throughout the period of incubation. At night they slept together with the eggs. Although the nests were built by monogamous pairs, with no other individuals taking an interest in them, after the eggs were laid additional black-faced birds attached themselves to them. At one nest, two black-faced Bushtits, identical in appearance, brought contributions of down. One, evidently the male of the pair, attached his tufts carefully, but the helper deposited his tufts so negligently that the former had to finish the slipshod work. Soon after incubation began in one of the nests, a black-faced helper slept in it for a night or two, but this was not continued. As a rule, only the two parents passed the nights with the eggs.

When, after fifteen days of incubation, the tiny downless nestlings hatched, the black-faced helpers joined the two parents in bringing them minute insects and green larvae. Each of two nests had a single helper, who not only fed but also brooded the nestlings. At the third nest, where at least three auxiliaries brought food, I was not certain that any of them brooded. At this nest, the five attendants (two parents with three assistants) fed four two-week-old nestlings 115 times in three hours one morning. The mother brought 32 meals, the four black-faces 83. Although feeding was now the adults' chief occupation, occasionally they brought still more down to pad the pouch. In return for their services, the helpers were permitted to sleep in the nests—no small reward on the chilly nights late in the dry season. In the best-attended nest, two parents plus two helpers slept with four feathered nestlings, probably packed together in layers. I wondered how the young Bushtits, at the bottom of the feathered mass, avoided suffocation in their thickly padded nest.

All the five or more helpers that I found had black faces. At only one nest did a single gray-faced female stranger appear. Mildly chased by the male parent, she was tolerated by the nestlings' mother and even spent some time in the pouch with her; but this stranger was not seen to bring food and did not stay.

At each of my three nests, all four eggs hatched and all four nestlings lived to fly. All twelve of them had black faces like their fathers, which puzzled me, as this suggested a sex ratio improbably unbalanced in favor of males. When the sexes of a passerine species differ in plumage, the young are usually more similar to their plain mothers than to their more colorful fathers. Accordingly, I assumed that all the helpers were yearling or older males, who could not find mates because of the paucity of females; and the taxonomic works that I consulted, based upon collected specimens that presumably had been properly sexed, confirmed this assumption. In more recent years, it has been demonstrated that, in some northern races of the Bushtit, juveniles of both sexes have black faces, which both sexes lose as they mature, or only the males retain. Whether this is true of the most southern populations of the species has not been investigated, as might be done with no sacrifice of life, simply by looking at enough feathered nestlings or dependent fledglings. If no gray-faced individuals appear among them, juveniles of both sexes are certainly black-faced. If, as I believe, some or all of the helpers that I watched, of whatever sex, were yearlings, we have the interesting phenomenon of one of the smallest of birds taking more than a year to acquire adult plumage, at least in the females—as happens in the males of some of the equally diminutive manakins.

It was so difficult to extract the nestling Bushtits without risk of injuring them that until they were feathered, I examined them by pushing back the flexible hood of the pouch and peering down through the doorway. When they were two weeks old, I tried for the first time to remove one for closer scrutiny by coaxing it upward in the flexible nest. While I worked at this, the parents and two helpers flitted close around me, protesting. After I had laboriously extracted a nestling, the other three emerged with a rush. Although fully feathered, they could fly only a few feet; and three were easily captured and replaced in the nest. The fourth hid so well that after an hour's search failed to disclose it, I reluctantly departed, leaving it in the open. Returning late in the afternoon, I found this truant trying to rejoin its siblings, while the attendants fed it generously. By hopping from twig to twig and flying short distances between them, it gained the pouch but could not reach the hooded doorway until I caught it and stuck its head inside.

From this experience, I confidently looked for the young, who left spontaneously when eighteen or nineteen days old, to return to their nests and sleep with their parents and helpers. In this I was mistaken. Although an admirable shelter during the last two months of the dry season when the Bushtits nested, after the rains returned in mid May the sodden pouch would hardly have made a comfortable or healthful dormitory. Now the Bushtits, young and old, joined in flocks which roosted in the crowns of trees in open woods or at the edge of woodlands, with naught but dripping foliage to shelter their tiny bodies from the cold mountain rains. Near the northern extremity of their range in the state of Washington, Bushtits roost with a short distance between individuals on milder winter nights; but when the temperature falls below the freezing point, they huddle in contact through the long hours of darkness.

These birds that clump together on frosty nights belong to one of the races formerly called Common Bushtits (*Psaltriparus minimus*), in which adults of both sexes lack black on their faces. From southwestern British Columbia, these races spread over western United States from western Wyoming and central Texas to the Pacific coast. In winter they live in stable flocks of up to fifty members, which defend their territory against neighboring flocks. As nesting time approaches, monogamous pairs scatter over the group's territory to raise their families. At some nests a third Bushtit, of undetermined sex, helps the parents incubate, brood, and feed the young. Or birds from a passing flock, usually males, may pause for a while to help a pair feed its nestlings. After the young can fly, they are led into the flock, where they are sometimes fed by males other than their fathers, who in turn may give food to fledglings not their own. These young of the year often permanently join their parents' flock, whose members tend to be closely related.

Long placed in the family Paridae with the titmice and chickadees, the Bushtit has recently been transferred to the Aegithalidae with the Long-tailed Tit (*Aegithalos caudatus*) of Eurasia. Although widely separated geographically, these two diminutive birds have much in common, in their flocking and foraging habits, in the ways they sleep, in their nests frequently attended by helpers. Five and a half inches (14 centimeters) in length, the Long-tailed Tit has a white head with a bold black stripe arching above either eye to join the predominantly black upper plumage, which is diversified by a pinkish patch on the shoulders and lower back and by white edges on the wing coverts. The underparts are white, and the three outer feathers of the strongly graduated tail have white margins and tips. Outside the breeding season, the tits wander through hedgerows, coppices, and woodlands in chattering flocks of half a dozen to about thirty individuals, incessantly searching leaf and bark for tiny insects. In winter they roost, low amid dense shrubbery or on the bare branch of a tree, pressed together in a row or, in the coldest weather, in the midst of a thicket or in a hole, cuddled together in a compact ball with their long tails sticking out at diverse angles.

Among brambles, in a shrub, or high in a tree, a male and female, working together, build an oval nest with a side entrance that has won for its makers the name "bottle tit." Of green moss, shredded wool, lichens, and cobwebs skillfully felted into a compact fabric, the 5-inch (13-centimeter) structure is usually covered with lichens which assimilate it into its setting. The interior is stuffed with feathers, of which over two thousand were counted in one nest. From the start of building to the beginning of laying the season's first eggs, from three weeks to over a month may elapse, but later nests are completed and occupied much more rapidly. The female usually lays from seven to ten eggs, but clutches of about twenty, presumably the product of two birds, have been reported. While the female incubates, her mate, who does not share this duty, brings food to her. He sleeps in the nest with her and helps feed the young.

At a substantial proportion of Long-tailed Tits' nests, helpers appear after the eggs hatch, rarely as early as the nestlings' fourth day and increasingly during the latter part of the nestling period. These auxiliaries are adults of both sexes, usually one to a nest but occasionally two, who contribute importantly to the nourishment of the nestlings, sometimes bringing food about as often as either parent or even more frequently. They hardly increase the rate at which the young are fed, because the parents themselves relax their efforts and bring less. Nests of Long-tailed Tits suffer very high predation, and these birds rarely lay replacement sets after the end of April in England. Many of the adults who are left early in the season without young of their own appear to serve at nests of more fortunate neighbors, who do not defend territories. These helpers continue to feed the young after they fledge. I have found no information as to whether, like Bushtits, they sleep in the nest with the nestlings and their parents. With hundreds or thousands of feathers, a brood of up to ten nestlings, and one or two long-tailed parents, the snug nest would appear to be overcrowded without these auxiliaries.

After the young hatch, parent Long-tailed Tits perform a hover display, flying steeply upward from a perch for 1 or 2 feet (30 to 61 centimeters) and hovering for about a second, before descending with a flutter to the same or another perch. Both sexes give this display, always near the nest, either before or after they feed the nestlings. Sometimes they display while holding a beakful of caterpillars or carrying a fecal sac. Why they should risk drawing a predator's attention to their nest by a conspicuous dis-

Brown Treecreeper

Long-tailed Tit

play is puzzling. The most probable explanation is that they perform so to advertise to whatever unemployed tits might be in sight that their brood has hatched and they would welcome assistance. Anthony Gaston, who described this display, saw no helpers come to a nest until after the parents had given it.

The Long-tailed Tit and the black-faced form of the Bushtit, residents in very different climates, present two contrasting types of cooperative breeding. As in many other tropical birds, including their neighbors the Banded-backed Wrens (*Campylorhynchus zonatus*), Bushtit helpers appear to be mostly yearlings with no previous experience at nests. Long-tailed Tit helpers seem to be chiefly adults who have lost their broods.

REFERENCES: Addicott 1938; Bent 1946; Ervin 1977, 1978; Gaston 1973; Lack and Lack 1958; Skutch 1935, 1960; Smith 1972.

30.

NUTHATCHES

Sittidae

The nuthatch family is widely spread over the eastern hemisphere, from the British Isles to Japan and the Philippines and southward into Africa and Australia, but only four of its twenty-seven species inhabit the New World, from Alaska and Canada to the highlands of central Mexico. Ranging in size from about 3.5 to 7.5 inches (9 to 19 centimeters), they are compactly built, short-necked birds with strong, sharp-clawed feet—which enable them to climb over trees, upward, downward, or sideways, without support from their short tails—and with slender, sharp bills for plucking insects and spiders from crevices in the bark. A few species climb over rocky cliffs or walls instead of trees. Most are rather plainly colored, often gray or subdued blue above, white to reddish brown below, but a few display bright blue or red. They nest in crannies in trees or rocks or in birdhouses, which they line with the most varied soft materials. If the entrance to the cavity is too wide, some species narrow it with a rim of clay, or they caulk gaps in the walls. They lay from three to ten white, spotted eggs, which are usually incubated by the female alone.

The bluish gray–and–white Pygmy Nuthatch (*Sitta pygmaea*) is widely distributed over western United States, chiefly in the mountains, where it forages high in tall pines. In a four-year study in central California, Robert Norris found three individuals in attendance at eight of thirty-six nests. The third bird was invariably a male, usually a yearling but sometimes older, who was not actually mated to the female of the monogamous pair. He helped to excavate the nest cavity and to line it with feathers, shreds of bark, moss, rabbit fur, wool, snakeskin, plant down, and various other materials. He fed the female during the sixteen days that she incubated her five to nine eggs. Then he helped the parents to nourish the nestlings with insects and spiders and to clean the nest. After the young emerged

Pygmy Nuthatch

at the age of about three weeks, the helper continued to attend them until, when about seven weeks old, they became self-supporting. Pygmy Nuthatches rear only one brood a year.

At all stages of the nesting, the extra male slept in the nest cavity with the parents and their eggs or young. After the breeding season and throughout the winter months, families of Pygmy Nuthatches tend to remain intact, and young and old sleep together in the nest hole or some other suitable cavity. In the cooler months, members of other families are not repelled from the lodging and may augment the number of sleepers to eleven. Many of the nest helpers are young males who have thus remained closely associated with their parents since the preceding breeding season. Their failure to rear families of their own is evidently caused by the paucity of females, since, for obscure reasons, the sex ratio is strongly unbalanced in favor of males.

In addition to the male auxiliary at a substantial proportion of the nests, Pygmy Nuthatches cooperate in other ways. Five or six of them may excavate a cavity, which is later occupied by a single pair. When two families forage together, an adult of one sometimes feeds a fledgling of the other family. At one nest, nine nestlings, not an abnormally large brood, were fed by four adults.

Among Brown-headed Nuthatches (*S. pusilla*) of southeastern United States, as among their western counterparts, males greatly outnumber females. In Georgia, Norris discovered trios at three of seventeen nests. All three helpers were males, who assisted at all stages of the nesting, but at night they did not join the parents in the nest cavity with the eggs or young. After fledging, the young do not return to sleep in the nest but are led to roost amid the foliage of a tree, while their parents sleep at a distance, perching side by side on a twig. Although at times, especially on cold nights, as many as four Brown-headed Nuthatches may sleep in a birdhouse, they occupy closed dormitories less consistently than Pygmy Nuthatches.

A pair of Brown-headed Nuthatches that lost their nest helped feed nestlings of a neighboring pair. At another nest in Georgia, two pairs cooperated to line a nest cavity; members of one pair fed the incubating female of the other pair; and all four adults fed the seven nestlings. Whether they were the progeny of one or both pairs was not known.

Cooperative breeding appears not to have been reported of the other two North American nuthatches, the White-breasted (*S. carolinensis*) and the Red-breasted (*S. canadensis*), or for any Eurasian or African member of the family. A European Nuthatch (*S. europea*), whose nest was only 3 feet (1 meter) away from a box occupied by Common Starlings (*Sturnus vulgaris*), often fed its neighbors' nestlings and removed their droppings.

In Australia the Orange-winged, or Varied, Sittella (*Daphoenositta chrysoptera*) forages singly or in parties of up to twelve individuals, among whom males predominate. They may breed in simple pairs or cooperative groups which frequently engage in mutual preening. Their cup-shaped nests, decorated with small strips of bark, are built in vertical forks by both members of a pair or by as many as four birds working closely together. At nests near Armidale in New South Wales, Richard Noske learned that only one female incubated the three bluish white, speckled eggs and brooded the nestlings. She was fed by three other adults, either while she covered the eggs or when she left the nest to meet the approaching group, scurrying over the branches with quivering wings and uttering a rapid, high-pitched chatter, like a begging fledgling, while she received food from her attendants. After the nestlings hatched, they were fed not only by both parents, who supplied over half of their meals, but also by a second adult male and two juveniles, a male and a female, who together brought food as often as the mother did. An adult female who belonged to this group of six was not seen to feed the nestlings, probably because their mother chased her from near the nest whenever she approached. All five of the sittellas who fed the nestlings also cleaned the nest.

After the young left the nest at the age of seventeen days, the same five birds continued to attend them, feeding them occasionally until they were about eighty days old. The whole group roosted on thin dead branches, not always in the same tree. The breeding male regularly arrived first, followed by other group members, who alighted near the tip of the branch and crept in a spiral course toward the base, passing beneath birds already present and squeezing upward between them. The females and young in the middle of the compact row were the first to tuck their heads back among the feathers of their shoulders and sleep. The adult males at the exposed ends remained awake and vigilant until after dark.

Each of two nests of Orange-winged Sittellas studied by S. Marchant had up to seven adult attendants. At these nests the incubation period was twenty days, and the young remained in the nest for another twenty days. Both of these periods are surprisingly long for a small bird with an open nest.

References: Bleitz 1951; Houck and Oliver 1954; Marchant 1984; Norris 1958; Noske 1980a; Powell 1946.

31.

TREECREEPERS

Climacteridae

The six species of treecreepers inhabit Australia, with one extending to New Guinea. They are small birds, about 6 inches (15 centimeters) long, gray-brown to rufous and black above, with a superciliary stripe, whitish or rufous wing bars, and pale, longitudinally streaked underparts. Although in appearance and habits they resemble the creepers (Certhiidae), they are not closely related. Their tail feathers are not pointed, as in northern creepers. Without this aid to climbing, they spiral upward around tree trunks while with sharp, slightly downcurved bills they probe the bark for insects and other small creatures. They also search for invertebrates on the ground, amid fallen trunks and branches. They utter shrill, high-pitched whistles. In hollows in trees or fallen logs they lay two or three, rarely one or four, white or pinkish eggs spotted with reddish brown. Both sexes attend the nest.

In open woodland at Swan Vale in New South Wales, Richard Noske studied a banded population of Brown Treecreepers (*Climacteris picumnus*) for three years. Here the birds lived in pairs, trios, or quartets, with more pairs in some years and more of the larger groups in others. As in other parts of eastern Australia, males were much more numerous than females, and with one probable exception, no group contained more than a single female, who was subordinate to all the males. Since the sexes were equally represented among nestlings and fledglings, the paucity of females in the adult population probably resulted from their earlier emigration or expulsion from their natal territories, while male offspring often remained for two years or more and helped their parents. In addition to the usual exploration of trunks and branches for food, Brown Treecreepers foraged much on the ground, where group members could alert one another to the approach of terrestrial reptilian and mammalian predators, especially the introduced fox. In contrast to many other cooperative breeders, treecreepers do not seek bodily contact with their companions, nor do they preen each other. At night they sleep singly in hollow trunks or branches of trees or in old nest cavities, sometimes high above the ground.

Brown Treecreepers breed in the austral spring or early summer. They nest in holes in trunks, in stubs of branches with hollow ends, usually on dead trees, or in cavities in stumps and fence posts. Nests are often a foot or two (30 to 61 centimeters) deep, with entrances 2.5 to 5 inches (6 to 13 centimeters) wide. At Swan Vale, all group members aided in nest construction, with the female being by far the most active in the first stage, when she carried in dry grass and bark for the foundation; in the second stage the breeding male and helpers brought most of the fur, feathers, thistledown, or snakeskin for the lining. During the incubation and most of the nestling periods, the males continued to add more of these materials. Only the female incubated the two or three eggs, taking sessions that ranged from two minutes to over an hour—most often less than thirty minutes—and spending from 50 to 80 percent of the birds' active day in the nest. The males in her group fed her both on and off the nest, the helpers together bringing food to her almost as often as her mate.

After an incubation period of sixteen or seventeen days, the nestlings hatched blind and naked, except for tufts of gray down on the head and back. The young treecreepers attained adult body weight when two weeks old, but like many hole nesters, they remained safely in their nests until they grew stronger. Here they were fed by the two parents and their helpers, at an average rate of eighteen times per hour. Three helpers contributed almost as many meals as the two parents together, and four brought well over half of the nestlings' food. All the attendants cleaned the nest. When, at twenty-five or twenty-six days of age, the fledglings left the nest, they sought the hollow ends of dead stubs, where for several days they called for food and were fed by parents and helpers. When fifty-five to sixty-five days old, the young became self-supporting. Brown Treecreepers are double-brooded, and occasionally juveniles from the first nest help feed siblings in their parents' second nest, as one male did when about seventy days old, and his sister did a few days later.

During most of the year, groups of Brown Treecreepers defend their territories even against related groups, but while feeding nestlings and fledglings they relax territorial defense. In these circumstances, a breeding male may alternate between feeding his parents' nestlings and his own, and nonbreeders commonly feed young at two nests, a situation rare among cooperative breeders but found also in wood-swallows, Chestnut-bellied Starlings (*Spreo pulcher*), San Blas Jays (*Cyanocorax sanblasianus*), and other jays. Thus, the number of attendants at a nest may be greater than the number of birds in a group.

A peculiar habit of Brown and other treecreepers is sweeping, which is performed by both sexes of the cooperative species but only by the female of the unsocial White-throated Treecreeper (*Climacteris leucophaea*). Holding some soft material in its bill, the bird rubs it sideways over the mouth of the nest hole and surrounding surfaces, including adjacent branches, sometimes continuing this for half an hour. Usually some clear, translucent material such as snakeskin, insect wings, or fragments of plastic bags is chosen for this operation, the purpose of which may be to obliterate the scent left by possums or other small animals, who compete with treecreepers for cavities in trees and sometimes ruin their nests. This activity is much like the bill sweeping of White-breasted Nuthatches (*Sitta carolinensis*), which—around the doorway of their nest cavity and over the surrounding trunk—rub tufts of fur or fragments of a plant, more often an insect such as a blister beetle, apparently as a chemical repellent to small animals that covet the nest hole or its contents.

Noske also studied in detail the Red-browed Treecreeper (*Climacteris erythrops*), which inhabits wet sclerophyll forests in southeastern Australia and lives in cooperating groups much like those of the Brown Treecreeper. Likewise, Rufous Treecreepers (*C. rufa*) and Black-tailed Treecreepers (*C. melanura*) breed both cooperatively and as simple pairs, as does the Brown Treecreeper. The other two species in the family, White-throated Treecreepers and apparently also White-browed Treecreepers (*C. affinis*), are at all seasons less sociable and nest without helpers. Instead of retaining their offspring on their territories, White-throated Treecreepers harass and evict them only thirty to forty-five days after they fledge.

The production of fledgling Brown Treecreepers per group was about 2.5 times as great as that per simple pair, but for each attendant adult, the difference between the productivity of pairs and groups was insignificant, 0.3 young per individual in pairs and 0.5 per individual in groups. Noske provides an interesting comparison between the productivity of noncooperative White-throated Treecreepers and Red-browed Treecreepers nesting in the same locality and between that of White-throated Treecreepers and Brown Treecreepers in another locality. Each of the two comparisons showed that pairs of White-throats produced substantially more fledglings per adult than did groups of the other species. Evidently, the advantages of cooperative breeding by treecreepers must be sought elsewhere than in their production of young.

REFERENCES: Noske 1980b, 1982.

32.

WRENS

Troglodytidae

With the exception of the Winter Wren (*Troglodytes troglodytes*), in Britain called simply the Wren, widely distributed in Eurasia as well as North America, all sixty species of the wren family are confined to the continents and islands of the western hemisphere. These small, sharp-billed, insectivorous birds, rarely as much as 8 inches (20 centimeters) long, are most abundant in the forests, thickets, grasslands, and marshes of the tropics, but some extend to fairly high latitudes north and south. Lacking bright spectral colors and sexual differences in plumage, they are clad in shades of brown, rufous, and gray, often diversified with black and white; many are conspicuously spotted or barred. More notable for their voices than for their plumage, many sing exquisitely, the male and female of a pair joining in antiphonal duets. If not in a hole or cranny, their nests are always roofed structures with a sideward- or downward-facing entrance. The nests are often built in considerable numbers and at all seasons, for many species sleep in them throughout the year as well as rear broods in them. Only the female incubates the two to, rarely, nine or ten eggs, but the male usually helps her feed the young.

Among the larger wrens is the 7- to 8-inch (18 to 20 centimeters) Banded-backed Wren (*Campylorhynchus zonatus*), boldly marked on back, wings,

Banded-backed Wren

and tail with black or dusky bars alternating with white, gray, or buff. Its white throat and breast are heavily spotted with black; its sides, flanks, and abdomen are tawny ochraceous. An adaptable species, it ranges from east central Mexico to northwestern Ecuador and from warm rain forests at sea level to coniferous woodlands on frosty heights at 10,000 feet (3,050 meters). Among the many places where I have found it are open woods of pine, oak, and alder high on Guatemalan mountains, pastures with scattered trees, and, at lower altitudes, the margins of wet forests, shady cacao plantations, and the palms around seaside cottages. As versatile in their ways of foraging as they are in their choice of habitats, Banded-backed Wrens hunt insects and spiders amid the foliage of trees, search among epiphytes for them, pull gray lichens from boughs to see what may be lurking beneath, creep over trunks and branches like overgrown nuthatches, and rummage through the ground litter.

Among the most social of wrens, Banded-backs live throughout the year in groups of usually six to twelve individuals, who dwell in amity and preen their companions. The territory of each group contains one to several globular nests, sometimes nearly a foot (30 centimeters) in diameter, with a wide doorway in the side. The thick roof and walls are composed of straws, weeds, pine needles, moss, lichens, sheep wool—whatever the locality affords for making a dry and cozy dormitory. In the highlands, the nests are often placed conspicuously in pine or oak trees; amid the rain forest they are frequently tucked out of sight amid heavy masses of epiphytes. Rarely as low as 6 feet (2 meters) up in a garden, they may be 100 feet (30 meters) high at the forest's edge. Building, which occurs at all seasons, is usually so desultory that it is difficult to learn how many members of a group participate in it, but I have watched at least four carrying materials to a nest. All members of a group sleep in such a nest, usually retiring early in the evening and, on cold and blustery mornings high in the mountains, emerging long after neighboring birds with less snug roosts have become active. Possibly without such adequate shelter Banded-backed Wrens could not thrive in such a wide diversity of climates.

The breeding season begins with much restlessness among group members, frequent changes of domicile, lively pursuits, and much dueting in harsh, rollicking voices by the one breeding pair that I found in each group. Finally, the female lays from three to five eggs, immaculate white or faintly spotted with pale brown, sometimes in an old, weathered, but sound dormitory nest rather than in a newly built structure. While incubating and brooding nestlings, she sleeps alone; other members of her group lodge together in another nest. At one nest, a single helper joined the father in feeding the young. Their mother brooded by day, in decreasing amounts until the young birds were about two weeks old, and thereafter by night as long as they remained in the nest, but she brought little if any food to them. Another nest was the center of interest of a group of seven, of whom at least three brought food to the young while another brooded them. I suspected that the whole group attended this nest, but since I did not succeed in marking these wrens, I could not prove this. The helpers at these nests were probably mostly yearlings.

After the young Banded-backs fledge at the age of eighteen or nineteen days, they are led in the evening to sleep in one of the group's nests. If this is difficult to reach by fledglings who have been practicing flight for only one day, an attendant instructs and encourages them by going in and out of the nest in their presence, if necessary many times until, after repeated fumbling attempts, they succeed in reaching a doorway with no convenient perch in front and pulling themselves inside. The adults may sleep with them or in a nearby nest. In the morning, the old wrens become active long before the newly emerged fledglings and may bring a little food to them before they in turn leave the dormitory. After the young fly more expertly, the whole group (with certain changes in membership) lodges together until the following breeding season. Some families change residences rather frequently, while others occupy the same dormitory for months together.

The pioneer study of Banded-backed Wrens that I made in the highlands of Guatemala half a century ago left unanswered a number of questions that have been addressed by more recent studies of cooperative breeders, using modern methods. As far as I know, nobody has since made a thorough study of this wren or of any of its relatives that live in groups throughout the year and apparently breed cooperatively. These include the Gray-barred Wren (*Campylorhynchus megalopterus*), Boucard's Wren (*C. jocosus*), and the Giant Wren (*C. chiapensis*), all of Mexico. In Costa Rica, Rufous-naped Wrens (*C. rufinucha*) not only lodge three or four together in pocketlike nests through much of the year, but I found three individuals greatly concerned about a nest with three eggs that I could not stay to study.

In Arizona, Cactus Wrens (*C. brunneicapillus*) sixty-six days old fed fledglings of their parents' next brood. Although juvenile helpers are occasional in this northernmost member of a tropical genus, it is not a cooperative breeder.

From southeastern Mexico throughout Central and South America at low and middle altitudes, in the Antilles and Falkland Islands, the small, faintly barred, brownish, blithely singing Southern House-Wren (*Troglodytes musculus*) nests in crannies in buildings and gardens as well as farther afield and is one of the most familiar of birds. Unlike the migratory, frequently polygamous Northern House-Wren (*T. aedon*), it resides permanently in monogamous pairs, of which the male helps build the nest and takes his full share in feeding and guiding the young. As the day ends, fledglings newly emerged from the nest are led by their parents to sleep in a sheltered nook, sometimes in the nest space itself. If the young continue to lodge in the nest space, the parents usually evict them about the time the following brood hatches. Sometimes, however, the juveniles refuse to depart and continue to sleep close to their mother while she broods the nestlings. In these circumstances, they may help feed their younger brothers and sisters. Of the many families of Southern House-Wrens that I have watched, this happened only in one during the second year of my study, when the young appeared to be exceptionally precocious. Two juveniles of the first brood attended nestlings of the second brood, and the single survivor of the second brood brought food to the third, beginning at the age of fifty-four days.

Northern House-Wren

While attending the second brood, the seventy-three-day-old female of the first brood tried to usurp the place of her mother, who vanquished her after a day of the fiercest and most prolonged fighting that I have witnessed among birds. The more docile male of the first brood continued to attend his younger siblings. These observations revealed some of the factors that lead to cooperative breeding and are necessary for its firm establishment. Young birds tend to cling so tenaciously to the familiar home of their childhood that they stubbornly resist expulsion, and they would more often feed younger siblings if permitted by the parents to associate closely with them. But unless they remain subordinate to the parents, trouble brews—as in any human household where children try to dominate their elders. Southern House-Wrens mature too quickly for cooperative breeding, which is appropriate for slowly maturing birds.

For the Northern House-Wren, I have found records of only interspecific helpers. While Black-headed Grosbeaks (*Pheucticus melanocephalus*) of both sexes sat brooding their nestlings in Colorado, a wren gave them food, of which they ate some and passed some to their young. After the latter fledged, the wren fed them directly. A few days later, this wren brought food to a family of House Sparrows (*Passer domesticus*). In Idaho, a male house-wren fed well-grown Northern Flickers (*Colaptes auratus*) in a hole in an aspen snag, only about 15 inches (38 centimeters) below another in which his mate incubated. In an observation period of 124 minutes, he brought food to the nestling flickers sixty-two times. The helper repeatedly darted aggressively toward the female flicker, but rarely attacked the male flicker. Neither member of the flicker pair disturbed the female wren. After his own nestlings hatched, the male wren divided his contributions of food between them and the flickers' brood. Soon his own progeny appeared to claim all his attention.

For at least six days, a Winter Wren fed nestling Townsend's Solitaires (*Myadestes townsendi*), with no friction between the helper and a parent when they arrived simultaneously at the nest. In Great Britain, this same species has repeatedly been observed acting as a helper. One fed two Spotted Flycatchers (*Muscicapa striata*) after they had left their nest, which was close to that of the Wren. For at least four days, another Wren fed nestling Great Tits (*Parus major*) while his mate incubated. Still another Wren brought food to nestling Coal Tits (*P. ater*), nearly ready to leave their nest box, and twice passed food to the parent tits. Winter Wrens have also fed young Blue Tits (*P. caeruleus*), Willow Warblers (*Phylloscopus trochilus*), and Linnets (*Carduelis cannabina*) and nourished a European Cuckoo (*Cuculus canorus*) that was attended by Dunnocks (*Prunella modularis*). A pair of Wrens adopted fledglings of their own kind who had been hatched and partly reared by a pair of Great Tits.

A male Carolina Wren (*Thryothorus ludovicianus*) whose mate was incubating in a nest box fed not only her but likewise young Great Crested Flycatchers (*Myiarchus crinitus*) in a neighboring box.

In captivity, a Long-billed Marsh Wren (*Cistothorus palustris*), about a month old and still being hand-fed, repeatedly begged food from its attendant and then carried it to a pair of younger fledglings who were being reared in the same room. One evening, as the room was slowly darkened, the helpful juvenile settled in an open nest that had been placed in a clump of grass and began a subdued rapid twittering, which it continued until four younger marsh wrens joined it in the nest.

Over much of tropical South America east of the Andes and in eastern Panama lives a bird that was formerly known as the Black-capped Mocking-thrush, classified with the mockingbirds and thrashers, but is now called the Black-capped Donacobius (*Donacobius atricapillus*) and included among the wrens. Nearly 9 inches (23 centimeters) long, it has a black head and hindneck, deep brown wings and back brightening to chestnut on the rump and upper tail coverts, and a black tail broadly tipped with white on all but the central feathers. Below it is buffy, with a patch of deep yellow bare skin on each side of the neck. Its golden eyes, gleaming brightly in a glossy black head, give the impression that nothing escapes their penetrating gaze. Its thin black bill, long tail, and slender body that slips easily through low, dense vegetation impart an aspect of streamlined grace. The sexes are similar.

A bird without close relatives, the donacobius inhabits swamps and marshes overgrown with grasses, reeds, and cattails as well as narrow bands of tall, dense grasses at the margins of streams, ponds, and lagoons in open country at low altitudes. In pairs or small family groups, it plucks insects, spiders, and other tiny invertebrates from the marsh vegetation or picks them from the surface of the water between the crowded stems. I could rarely approach a small pond in northern Venezuela without being greeted by an alarmingly sudden and loud outburst of grating, rasping, or churring notes from the pair that most of the time lurked unseen amid the tall marginal grasses. Other utterances of these birds sounded like *cheeo cheeo cheeo cheeo cheeo* and *chu chu chu chu chu*, all strong and suggestive of excitement rather than melodious. In the several months that I studied these birds, I never heard them imitate anything. Neither in voice nor in behavior did they remind me of thrushes.

The display of a mated pair was a unique performance. Perching close together, the two partners spread their long tails fanwise, revealing a black central band broadly bordered with white, and wagged them from side to side while they opened their bills to emit contrasting notes. To the accompaniment of the female's prolonged sizzling or grating sounds, the male voiced a loud, ringing, liquid *who-it who-it who-it*. When two pairs met in a territorial dispute, they rested a few yards apart and displayed in this fashion.

Attaching her nest to broad blades of grass 2 feet (61 centimeters) above the watery mud at the pond's edge, the female donacobius built at a leisurely pace, with some help from her mate. The bulky open cup of narrow strips of grass leaves and fibrous materials differed from the nests of all undoubted wrens, which are either roofed or placed in a hole or sheltering cranny. Here she laid three lovely eggs, so densely mottled with light reddish brown that without close examination they appeared to be uniformly colored. She alone incubated, hatching three pink-skinned, downless nestlings after a period of seventeen days. During the additional seventeen days that the young remained in the nest, both parents fed them with insects and spiders, carried conspicuously in the tips of their slender bills, one at a time. Whenever I visited the nest, the watchful parents, coming near me, protested loudly with harsh rasping and churring notes.

The pair of Black-capped Donacobiuses that I watched in Venezuela, in a locality where the birds were rare, lacked helpers. Along the marshy shore of Cocha Cashu, an oxbow lake in Manu National Park in southeastern Peru, Richard Kiltie and John Fitzpatrick found eighteen territorial groups, each of two to four grown birds. The larger groups consisted of a mated pair with one or two recent young, who participated in territorial displays, nest surveillance, and feeding nestlings. In contrast to the Venezuelan nest, none of these Peruvian nests contained more than two eggs. Details of the helpers' contributions to nestling care are not available, but three of the four nests with a helper yielded two fledglings, while none of the ten nests attended by unaided parents produced more than one fledgling. The annual survival rate of breeding adults was 71 percent.

References: Anderson and Anderson 1962; Armstrong 1955; Bent 1948; Betts 1958; Kale 1962; Kiltie and Fitzpatrick 1984; Laskey 1948; Robinson 1962; Royall and Pillmore 1968; Selander 1964; Skutch 1935, 1953b, 1960, 1983b.

33.

GNATCATCHERS AND OLD WORLD WARBLERS

Sylviidae

Ornithologists differ confusingly about the classification of the multitude of small insectivorous birds sometimes included in the family Sylviidae. A recent tendency has been to reduce it to a subfamily, the Sylviinae, of the family Muscicapidae, the enormous assemblage of Old World flycatchers, thrushes, warblers, and their allies. For convenience, we here consider together the Old World warblers, the kinglets, and the gnatcatchers but not the wren-warblers, which are treated in chapter 34. Most of the approximately 340 species in the family are spread widely over Eurasia, Africa, and neighboring islands, with only the kinglets, the gnatcatchers, and the gnatwrens, 15 species in all, in the western hemisphere.

Rarely as much as 7 inches (18 centimeters) long, warblers and gnatcatchers are slender-billed birds that flit through woodlands, thickets, marshes, and savannas, restlessly seeking small insects and spiders. With rarely more than a touch of bright color, they are clad in olives, browns, grays, black, and white, sometimes streaked on back or breast. Many species are so similar in their severely plain attire that they are most readily distinguished by their voices—some sing charmingly. Their nests, in trees or shrubs or amid grasses or reeds, are open cups, roofed structures with a side entrance, deep purses, or pensile pouches, or they may be hidden in green leaves that tailorbirds stitch together. They lay from two to ten eggs of varied coloration, plain or spotted, that are incubated by both sexes or the female alone.

The Tropical Gnatcatcher (*Polioptila plumbea*), a tiny, slender bird widespread in tropical American woodlands and shady clearings, is blue-gray above and white below, with a long, black, white-bordered tail. Usually both sexes build the dainty, compact cup of fine fibrous materials, encrusted on the outside with lichens or fragments of green moss, liverworts, or algae. However, while a black-capped male built in a tree in a Costa Rican pasture, his gray-headed mate, diverted by the appeal of two nestling Golden-masked Tanagers (*Tangara larvata*) a few yards away in the same tree, quite neglected to help him. Until the young tanagers left their mossy open nest twelve days after I found her there, she continued to be engrossed by them.

Tropical Gnatcatcher

While the parent tanagers, coming and going together, brought mostly billfuls of fruits, the gnatcatcher diversified their diet with tiny insects, feeding them about as often as either of the parents did. She helped clean the nest, and she tried to brood the nestlings but apparently was not comfortable in a nest or covering nestlings too big for so diminutive a bird. Far from becoming friendly with the parents whose offspring she fostered, she tried to keep them away. If a tanager arrived while she sat in the nest, she jumped out and flitted around it with her tail spread to display the white outer feathers contrasting with the dark central ones, or she darted at the parent with drooping wings. At first the tanagers paid little attention to their uninvited helper, but as days passed and her attendance degenerated while she became increasingly annoying, they retaliated by flying at her more often. I witnessed several lively chases, in some of which the male gnatcatcher joined, apparently more to defend the vicinity of his own nest than because he was interested in the activity at the tanagers' nest, which I never saw him visit. Even after the fledgling tanagers flew away, the female gnatcatcher failed to use the nest her mate had built.

In Argentina, juvenile Masked Gnatcatchers (*Polioptila dumicola*) of the season's last brood stay with their parents for six or seven months, and subordinate individuals preen the dominant ones, mostly the breeding males. At one of twenty-six nests, a bigamous male and two females built, then incubated a double set of six eggs.

In Africa, at least three species of warblers have been found with helpers. In several different family parties of the Green-backed Warbler (*Eremomela pusilla*), more than two adults fed fledg-

Tit Hylia

lings. The average size of seventy-five groups of Dusky-faced Warblers (*E. scotops*) counted by the Vernons in Zimbabwe (formerly Rhodesia) was four birds, with extremes of two and eight. In permanent territories, flock members foraged together amid the foliage and twigs of the woodland canopy, with no antagonism even when only an inch or two apart. In sessions of simultaneous preening, one sometimes groomed another. They bathed together in treetop foliage wet with rain or dew. At night, flock members huddled in a row on a twig. In two flocks of four and five birds, at least three fed the fledglings, as did all members of another flock of four. Juveniles apparently remain in their natal groups and serve as helpers in the following breeding season.

The third African warbler known to have helpers is the Tit Hylia (*Pholidornis rushiae*). In Angola, Carl Vernon watched a family party of seven birds, of which two were fledglings. At least four adults, possibly five, fed the young in quick succession.

Both members of another African warbler pair, the Karoo Prinia (*Prinia maculosa*), fed nestling Layard's Tit-babblers (*Parisoma layardi*) in a nest near their own.

References: Fraga 1979; L. G. Grimes 1976; Martin 1968; Skutch 1960; Vernon and Vernon 1978.

34.

WREN-WARBLERS

Maluridae

Some eighty species of wren-warblers inhabit Australia, New Guinea, New Zealand, and neighboring islands. Sometimes lumped with the Old World warblers (Sylviidae) or even in a more inclusive assemblage of insectivorous birds, the Muscicapidae, here it seems more convenient to follow long-established usage and treat them as a separate family. Wren-warblers, also called Australian warblers, include some of the most charming and curious of small birds. Most elegant are the thirteen species of blue wrens (*Malurus*), affectionately called fairy wrens, of Australia—dainty, small-billed, long-legged birds that in family groups search for insects on or near the ground in bushy places. Adult males are blue, azure, purple, and black; some display expanses of red. Females are brownish and whitish. All carry their long, slender tails almost erect, like a few of the true wrens (Troglodytidae). This habit is also pronounced in the strange little brownish, streaked emu wrens (*Stipiturus*), whose tails, longer than their bodies, are composed of only six feathers, all delicately fringed with loose, lacy barbs. Wren-warblers build covered nests with a side entrance, in some species pendent, on or near the ground or high in trees; they lay two to five white eggs variously marked with shades of brown.

One of the earliest long-continued, detailed investigations of a cooperatively breeding bird was Ian Rowley's five-year study of Superb Blue Wrens (*Malurus cyaneus*) near the Australian capital, Canberra. These birds are among the most familiar of the blue wrens, for they inhabit the southeastern seaboard where most Australians live, and they have adapted well to suburban areas and city parks. The wrens live throughout the year in sedentary, territory-holding groups that may contain from six to a dozen or more individuals. Males are more numerous than females because they rarely leave their natal area, and then only in a preformed group, whereas young females, as in other cooperative breeders, more often emigrate and probably for this reason suffer higher mortality. In autumn, adult males molt into a brownish eclipse plumage much like that of females and juveniles. As the next breeding season approaches, the dominant male of a group regains his splendid nuptial attire about a month earlier than subordinate adult males, possibly because the latter are somehow inhibited by their inferior status.

Superb Blue Wrens sing sweetly, with a volume of sound surprising for such small birds. At dawn neighbors join their voices in a chorus that may continue for over half an hour. Females also sing and duet with their mates. Males have the unexplained habit of carrying yellow flowers or leaves. Mutual preening by all group members is frequent. To roost, the wrens line up in contact on a horizontal branch amid the dense foliage of a tree or shrub.

Low in a shrub or tussock of grass or well up in a tree, the female blue wren, usually working alone, builds of grasses, rootlets, and fine twigs, all bound together with spider webs and egg cases, her roofed nest with a side entrance. She lines it with wool, feathers, animal hair, or string. Although she often finishes her nest in three or four days, five days to a month may elapse before she lays her three or four eggs, which she alone incubates for thirteen to fifteen days. After the blind, naked, pink-skinned nestlings hatch, all self-supporting members of the family group bring them food, juveniles sometimes beginning when fifty-eight days old. The helpers most often include a single male in adult plumage, sometimes a male and a female, rarely two males and a female, with the parents making five adult attendants. Nests with helpers are, however, in the minority; at most only the two parents are present until juveniles of early broods assist with later broods. Rates of feeding are quite variable, sometimes reaching twenty-nine meals per hour by three attendants.

As she incubated alone, so the female parent broods alone, often continuing to cover the nestlings by night until, usually at the age of twelve or thirteen days, they leave the nest. Then, like many small birds, they enter a cryptic stage, hiding for a week or so amid dense, low vegetation. After they fly fairly well, they emerge into the open, where the parents and helpers continue to feed them until they are six or seven weeks old. At first their mother contributes to their support, but a week after they fledge she may start another nest, and soon she is fully occupied incubating her next clutch of eggs, leaving the still dependent fledglings in the care of other group members.

If all goes well, Superb Blue Wrens may raise two or three broods in a breeding season of six months. The second or third brood may hatch before the young of the preceding brood become self-supporting. Then, in an unassisted pair, the father must divide his attention between his nestlings and fledglings, who will probably follow him to the nest, clamoring for food and perhaps attracting predators. When a helper is present, it often takes full charge of the fledg-

Superb Blue Wrens

lings, keeping them at a distance from the nest and leaving the parents free to devote themselves to the nestlings. If the male of a pair dies, a subordinate male may inherit his territory, mate, and parental obligations; or sometimes a male from a neighboring territory, a bachelor or a subordinate, arrives to help care for nestlings or fledglings.

Parents seem unable to identify their own newly fledged young, as was shown when two neighboring broods hatched at the same time and the two sets of young in the cryptic stage intermingled. Territorial boundaries were ignored while both parental pairs, with the help of a bachelor from a third territory, fed the fledglings without discrimination. After about a week, one pair returned to its original territory and renested, leaving the six young in charge of the other pair and their assistant, who together reared them successfully.

Of 240 eggs laid in four years during Rowley's study, 158 hatched and 129 young fledged, giving an egg-to-fledgling success of 53.8 percent. Only 69 or slightly over half of these fledglings survived to independence, and only 29 of them (18.4 percent of the hatchlings) lived until the beginning of the following breeding season. The value of the helpers was attested by the fact that, in three seasons, 12 breeding groups with a total of 37 adult birds fledged 69 young, 1.9 per adult; 16 pairs, or 32 adults, fledged only 39 young, 1.2 per adult. In terms of nests, the difference is still more impressive. Each category had 32 nests; those attended by helpers yielded 2.2 fledglings per nest, whereas nests without helpers produced only 1.2. This difference cannot be attributed to the fact that pairs with helpers were more experienced as breeders than those without, since experienced simple pairs failed to produce, on the average, more fledglings than inexperienced simple pairs.

After moving from eastern Australia, where he watched Superb Blue Wrens, to the far west, Ian Rowley undertook an even longer study of the Splendid Wren (*Malurus splendens*), a rather similar species widely spread over the southern half of the continent to where it meets the eastern species along the Murray and Lachlan rivers. During seven breeding seasons, he banded and watched 214 Splendid Wrens in arid heathland on the steep slopes of Gooseberry Hill, inland from Fremantle. Here they were permanently resident on stable territories that averaged about 3.8 acres (1.5 hectares), large for such a small, weakly flying bird. The area of the territory was not correlated with the size of the group that occupied it. Males were more numerous than females in the ratio of about three to two.

Although only one-third of the breeding groups of the Superb Blue Wren consisted of more than a simple pair, nearly two-thirds of those of the Splendid Wren were larger. Most often they contained two adult males and a female, but a few had three males and one or two females, one had four males and two females, and one consisted of three adults of each sex. As in the Superb Blue Wren, the female built her nest with a side entrance and incubated her three eggs for thirteen or fourteen days, unhelped and unfed by her associates, whose aid at this stage was limited to sentry duty. After the young hatched, the father and his helpers became more active and brought them a wide variety of foods, from scale insects to such large items as grasshoppers and even small lizards. At the age of about forty days the young were mainly self-supporting, and when sixty days old they fed their siblings of the following brood, at some nests bringing the number of attendants up to seven—including parents and adult and juvenile helpers—and contributing about a third of the nestlings' meals. Juveniles as well as adults cleaned the nest. The young helpers worked much harder than was necessary to support themselves, but the experience of exerting themselves strenuously to find a wide variety of foods at a time of abundance would stand them in good stead in the leaner months after the breeding season.

Another service performed by both adult and juvenile helpers Rowley called herding. Each herder served as "nursemaid" for one younger juvenile, whom it led to the safest or most productive foraging areas, then acted as sentry while the youngster ate. Moreover, it helped its charge learn the boundaries of the group's territory, and it kept the young bird at a distance from the nest where its mother incubated a later set of eggs.

Helpers, mostly offspring of the parents whom they aided, sometimes continued in this role for as long as five years. The alternative of emigrating from their natal territory would have been for most a suicidal course, for the limited areas capable of supporting the wrens were already well tenanted and the emigrants would have been most unlikely to find a refuge. By staying at home, the male helper of highest rank had a 41 percent chance of attaining breeding status in any one year, either by replacing the lost breeding male of his own group or by finding an opening in an adjoining territory. A male who rose to alpha status in a territory would have no difficulty attracting a mate from among the female helpers.

These female auxiliaries were even more likely to rise to the status of breeders, if they survived, because of the higher annual mortality of their sex—57 percent as opposed to 29 percent for males. Their frequent deaths appeared to result less from predation at the nest than from the physiological strain and exhaustion of building the nest, laying three or four sets of eggs which together equaled their own weight of 0.35 ounce (10 grams), and incubating them without being fed, all of which, plus the energy-consuming molt that followed the breeding season, left them weakened and vulnerable to predation. Doubtless female blue wrens would survive longer if their mates and helpers did not limit their services to sentry duty and feeding the young but aided in more diverse ways, in the manner of many cooperative breeders. On the positive side, neither an adult nor a juvenile helper was ever known to be harmful, as by getting in the way, clumsily breaking an egg, or injuring a nestling. Never was more than one female found laying in the same nest.

Fifty-six percent of the Splendid Wrens' eggs produced fledglings. The greatest nestling mortality was caused by parasitic cuckoos, whose nestlings evicted the legitimate offspring from the nest. Despite the substantial aid given by helpers, the productivity of groups with them did not differ significantly from that of simple, unassisted pairs; and this was true whether fledglings or yearlings were counted. The helpers' contribution to reproduction was most clearly evident in the case of young, inexperienced females nesting for the first time. Without helpers, such novices raised less than a quarter as many fledglings per nest as experienced females did, but with helpers they produced two-thirds as many. Likewise, females who nested with helpers tended to live longer and leave more progeny than those who nested without assistance, but the difference was not statistically significant. One breeding male lived for at least nine years, one breeding female for no less than six years. By staying on their natal territory in cooperating groups instead of hazardously dispersing at an early age, Splendid Wrens prolong their life expectancy and form an experienced reserve to rebuild the population in the event of a catastrophic reduction, and they greatly lighten the burden of breeding.

Six other less thoroughly studied species of *Malurus* are known to breed cooperatively. Another five species live in groups, making it probable that all thirteen species of Australian blue wrens will eventually be found to have helpers at the nest. Moreover, such auxiliaries appear to be present in several other genera of the family.

Much more plainly attired than the blue wrens, little Yellow-tailed Thornbills (*Acanthiza chrysorrhoa*) of the Australian scrub build much larger nests. The nucleus of a nest, a dome-shaped structure with a side entrance, is an untidy accumulation of vegetable materials which may become 9 inches long by 12 inches wide (23 by 30 centimeters). Above this are one or more cup-shaped additions, which may double the size of the structure. The enclosed section is lined with feathers and vegetable down for the reception of the eggs; the cups remain unlined and their purpose is puzzling, as they are not used as dormitories. After the fledging of each brood, the nest compartment, if in good condition, may be refurbished for another set of eggs, or a new enclosed chamber may be built beside or beneath the original one. This sequence may be repeated for several years, until the structure becomes very bulky.

Both sexes build the compartment for the eggs and nestlings; the male, who continues to build while his mate incubates and even while nestlings are being fed, is chiefly responsible for the cups. For eighteen to twenty days, the female incubates the three to five eggs, unassisted and unfed by her mate, who, however, helps feed the nestlings during the seventeen to nineteen days that they remain in the nest. Julian Ford and others have learned that juveniles of one brood aid their parents in feeding nestlings and fledglings of a later brood. Other adults, who may be the mated pair's progeny of a previous year or possibly unmated or unoccupied individuals of other families, likewise assist in nourishing the young.

For four seasons, the Browns watched Yellow-tailed Thornbills nest in an olive tree in view of their kitchen window at Manjimup in southwest Australia. The young from each of the first nine broods were tolerated and fed in the nesting area while the female incubated the following clutch, but when these eggs hatched the juveniles were chased from the garden. None was seen to attend their younger siblings while in the nest; whether they fed the latter after fledging was not known. Finally, a juvenile of the tenth brood, hatched in January, was seen carrying materials to the nest that its parents were building in June. At first this helper brought small pieces and pushed them at random into the fabric, but with practice it became more expert, bringing more adequate materials more often and weaving them competently around the vertical twigs. The juvenile helped its father on the open cup atop the nest, while its mother worked alone in the brood chamber. After the mother started to incubate her three eggs, the young helper and its father continued to add to the false nest above her, until by the time the nestlings hatched it was almost closed at the top.

When the adult male started to feed the nestlings, the young helper flew to and fro with him but brought nothing. By the third day after hatching, the youngster came with very small items that it failed to deliver to the nestlings. As when it built, the juvenile improved with practice until, before the nestlings flew, it equaled the parents in the size and number of meals that it brought. It continued to feed its siblings after they fledged, and it also assisted the parents during two more nestings, the last of which yielded only a single Shining Bronze Cuckoo (*Chrysococcyx lucidus*). In this family of Yellow-tailed Thornbills, nest helpers were exceptional; as was true of the first nine broods, none of the young from the two subsequent broods was a helper.

Like the young Brown Jays (*Cyanocorax morio*) in the Costa Rican highlands, the juvenile thornbill developed competence in activities necessary for successful nesting while it helped its parents. Since it built only at the cup-shaped false nest, its work there contributed nothing to its parents' reproductive effort, but it suggested a use for this puzzling annex to the nest chamber—as a structure at which young thornbills could serve an apprenticeship in the art of nest building.

Another cooperative breeder among the wren-warblers is the White-browed Scrub-Wren (*Sericornis frontalis*), of which Harry Bell watched four adults feed three newly fledged young. He often found these small birds in groups of two to seven individuals, who tended to stay together from year to year. Other scrub-wrens who live in groups of three to five throughout the year probably breed cooperatively.

References: Bell 1983; R. J. Brown and M. N. Brown 1982; Cayley 1949; Ford 1963; Rowley 1957, 1965, 1975, 1976, 1981a, 1981b; Warham 1954.

35.

OLD WORLD FLYCATCHERS

Muscicapidae

Although thrushes, babblers, Old World warblers, and wren-warblers are now often treated as subfamilies of this family, in this book each of these groups receives a separate chapter. Here we shall consider only the four subfamilies formerly included in the Muscicapidae: the "typical" flycatchers of the subfamily Muscicapinae, the monarch and paradise flycatchers of the Monarchinae, the fantails of the Rhipidurinae, and the thick-heads or whistlers of the Pachycephalinae, which together number nearly four hundred species, widespread in Eurasia, Africa, Australia, and islands of the Pacific Ocean but absent from the New World, except for a few accidental occurrences.

In size these birds range from 3 to 21 inches (8 to 53 centimeters), including the very long tails of the paradise flycatchers. Such a large assemblage displays almost the whole gamut of avian coloration, from modest browns and grays, black, and white to lovely blues, reds, and yellows. In addition to their long, graceful tails, paradise flycatchers (*Terpsiphone*) wear high-peaked crests. Some flycatchers have facial wattles. With short, rather flat bills surrounded by bristles, Old World flycatchers catch insects in flight, pluck them from foliage, or seize them by a rapid descent to the ground. A minority are excellent songsters. Their nests, built by both sexes or by the female alone, are usually open cups, placed in a tree or shrub, in a cavity in a tree or cliff, or in a birdhouse. Their one to nine eggs, white, bluish, yellowish, or reddish, more or less spotted or mottled, may be incubated by both sexes of even the more ornate species. Both parents usually attend the young.

A number of African and Australian flycatchers at least occasionally breed cooperatively, but they await more detailed studies. The Forest Flycatcher (*Fraseria ocreata*) lives in groups of up to ten birds. At each of two nests in the Gabon, C. Erard watched three adults feed nestlings. The same observer found three individuals of another gregarious bird of African forests, the Chestnut-capped Flycatcher (*Erythrocercus mccalli*), feeding three nestlings. He also watched four Blue Flycatchers (*Trochocercus longicauda*) feed two nestlings. A pair of White-spotted Wattle-eyes (*Platysteira tonsa*) building a nest was assisted by a young female of an earlier brood. In a family group of three wattle-eyes, two juveniles from an earlier nest helped the parents feed two fledglings. Two adult Pale Flycatchers (*Bradornis pallidus*) fed a third who was brooding nestlings, and three adult Abyssinian Slaty Flycatchers (*Melaenornis chocolatinus*) brought food to a brooding individual.

White-spotted Wattle-eye

In New South Wales, Australia, Crested Shrike-tits (*Falcunculus frontatus*) remain in groups of three to five

Crested Shrike-tit

individuals long after the breeding season. Two fledglings were repeatedly fed by two adult females and a male. In another family group, two adult males and one female fed two recently fledged young. Two female Hooded Robins (*Petroica cucullata*) took turns incubating in the same nest. Of three successive nests of this flycatcher, two contained four eggs, which could be separated by appearance into two pairs that had apparently been laid by different females. Only one male was seen at these nests.

In western Australia, the Browns watched three nests of White-breasted Robins (*Eopsaltria georgiana*) at which, respectively, four, four, and two color-banded birds helped the female parent to feed nestlings. At two nests of the

Pied Flycatcher

Western Yellow Robin (*E. griseogularis*), three and four adults fed the young.

The black-and-white Pied Flycatcher (*Ficedula hypoleuca*) of Europe and western Asia nests in tree cavities and birdhouses and is often polygynous. Pairs feeding their young are sometimes helped by a stranger whom they do not try to drive away. While a lone male was attending nestlings who had lost their mother, a new female arrived and, despite his mild attacks, joined him in feeding them. When strange young Pied Flycatchers, already self-supporting, rested atop an experimental nest box in which hungry nestlings called for food but were inaccessible to their parents, the latter sometimes fed them. A pair of Spotted Flycatchers (*Muscicapa striata*) who lost their nest fed Eurasian Blackbirds (*Turdus merula*) in a neighboring nest.

REFERENCES: R. J. and M. N. Brown 1980; L. G. Grimes 1976; Haartman 1953, 1956; Howe and Noske 1980; Rowley 1976; J. Southern 1952.

36.

THRUSHES

Turdidae

Some of the 303 species of the thrush family are known to almost everyone interested in nature, for they have established themselves over the whole Earth except the polar regions and some remote islands, and where not native they have been introduced by admirers, as in New Zealand. Many are familiar inhabitants of gardens and parks. From 5 to 13 inches (13 to 33 centimeters) long, they may be plainly clad, as in many species of the genus *Turdus*, or colorful with red, orange, or blue. In addition to much fruit, they eat insects and other invertebrates, which many of them gather as they hop over the ground, tossing leaves aside with their bills. Their nests—with few exceptions open cups of vegetable materials, often reinforced with a layer of hardened mud—are situated amid the foliage of shrubs or trees, on the ground, in crevices among rocks, holes in trees, or birdhouses, or rarely in underground burrows. Their two to six or, rarely, more eggs, often blue, blue-green, or otherwise tinted, plain or with markings, are nearly always incubated by the female only, but the male helps feed the young.

This family includes many of the most brilliant and loved of songsters. Their songfulness is frequently associated with the strong territoriality of monogamous males who repel all individuals of their species except mates and dependent young—behavior hardly compatible with cooperative breeding. Nevertheless, thrushes occasionally help at nests of their own or different species. (Although usually given familial status in the past, in some recent classifications the thrushes have been reduced to a subfamily, the Turdinae, of the huge complex of Old World insectivores, the Muscicapidae.)

After rearing two of her own young, a female Eurasian Blackbird (*Turdus merula*) continued for two or three weeks to offer food to any bird that came near, and an adult European Robin (*Erithacus rubecula*) was among those who accepted. On two occasions, a fledgling blackbird, who strayed rather far from its nest on the first day after leaving it, entered the territory of another pair of blackbirds who had fledged young. In both cases, the male of the new pair promptly adopted the wanderer and fed it until it could care for itself; it was never seen to receive food from its own parents. This was unusual, as young that beg from strange adults commonly receive nothing. When two families mingle, the parents have always been seen to feed only their own fledglings.

European Robin

As among blackbirds, fledgling American Robins (*Turdus migratorius*) that become separated from their parents are occasionally adopted by other adults. In aviaries, young robins fed still younger birds of other species. Robins sometimes share a nest with another individual, of the same or a different species. The most curious of a number of instances that have come to my attention is that of a robin and a Mourning Dove (*Zenaida macrura*), each of whom laid two eggs in the robin's nest. They took turns incubating, then fed and brooded the nestlings until they were eight days old. On the following day, the four nestlings died. A robin and a Gray Catbird (*Dumetella carolinensis*) each built a nest in the same clump of lilacs. Both took turns incubating the catbird's eggs, and when the young hatched they were brooded by both the robin and the catbird. Instances of cooperation by robins and House Finches (*Carpodacus mexicanus*) are given in chapter 49 on grosbeaks, finches, and sparrows.

Nine days after the first brood of a Wood Thrush (*Hylocichla mustelina*) left the nest, their mother fed a strange young bird who, with its parents, had entered her territory. In an aviary, an old, unmated Wood Thrush helped feed fifteen nestlings of several species, including Wood Thrushes, Veeries (*Catharus fuscescens*), Bobolinks (*Dolichonyx oryzivorus*), Northern Cardinals (*Cardinalis cardinalis*), and orioles (*Icterus*). Her coworker was the young Eastern Bluebird (*Sialia sialis*) mentioned beyond. A Swainson's Thrush (*Catharus ustulata*) assisted in feeding American Robins, bringing food to their nest at least twelve times in four hours, while the parent robins were present. Three adults of the Gray-cheeked Thrush (*C. minimus*) fed the young in one nest.

In all three species of North American bluebirds, the Eastern, the Mountain

American Robin

(*Sialia currucoides*), and the Western (*S. mexicana*), juveniles of an early brood sometimes feed their siblings of a later brood in the same year. In one such episode, five Eastern Bluebirds, all less than two months old, diligently cared for the four nestlings of the second brood, beginning when the latter were three days old. These young helpers also cleaned the nest. Occasionally juvenile Eastern Bluebirds help build a nest. More rarely, a yearling Eastern Bluebird helps its parents. In one recorded instance, a year-old male, after rearing a brood of his own, returned to his birthplace and assisted his parents to nourish their second brood, bringing food as often as the father. Two juveniles of the first brood were also helping at this nest, which yielded the rather unusual number of five fledglings. It is significant that these bluebirds had overwintered in their nesting area in Michigan instead of migrating. A female Eastern Bluebird incubated a set of eggs laid by another female who had been killed by a cat.

An adult male Eastern Bluebird fed nestling Northern House-Wrens (*Troglodytes aedon*), upsetting their parents, until his mate hatched young bluebirds, when he transferred his attention to his own offspring. In Florida, a pair of Eastern Bluebirds fed five young Northern Mockingbirds (*Mimus polyglottos*) against parental opposition, both before and after they left the nest. After a brood of nestling Eastern Bluebirds lost their father, a new male arrived and fed them, but much less than a male parent usually does. He also gave food to their mother. A six-week-old female Eastern Bluebird helped feed the fifteen nestlings of several species that were also nourished by the old Wood Thrush mentioned above, all of whom were being hand-reared in the same aviary. When

Northern Wheatear

slightly older, the same bluebird helped feed and brood nestlings of her own species, sometimes sitting in the nest beside their mother.

At two widely separated nests, young Northern Wheatears (*Oenanthe oenanthe*) helped their parents nourish second broods. On Baffin Island, a fully adult male, somewhat paler than the male parent, aided a mated pair feeding seven nestlings. Once their father mildly chased his assistant. While a photographer was filming a nest of Abyssinian Black Wheatears (*O. lugens*) in Africa, at least three adults, a female and two males, fed the nestlings, and three other adults of both sexes approached within 60 feet (18 meters) of the nest without provoking hostile behavior.

It appears that the more closely associated with humans a bird is, the more often it is seen helping other birds of various kinds in diverse situations. For his delightful book, *The Life of the Robin*, David Lack collected a number of instances of helpful behavior by this favorite of the British. Parents of fledglings occasionally feed strange fledglings, including those that differ considerably in age from their own. European Robins who had lost their own young fed a brood of nestling Song Thrushes (*Turdus ericetorum*). Other robins have ministered to young Song Thrushes, Eurasian Blackbirds, and, in several instances, Winter Wrens (*Troglodytes troglodytes*). A robin and a Willow Warbler (*Phylloscopus trochilus*) sat together on six eggs in a nest built by the warbler. A pair of robins raised a combined brood with a pair of Pied Wagtails (*Motacilla alba*). When two male robins were placed in the same cage, they fought until one broke his leg, after which his rival fed him. A story told by Eckermann in his "Conversations with Goethe" has been frequently repeated. Eckermann found two newly fledged Winter Wrens and wrapped them in a handkerchief to take home, but they escaped while he was passing through a wood. Three days later, he discovered them in a robins' nest, where they were being fed together with the nestling robins. Apparently the wrens, when seeking a snug lodging for the night, as is their habit, entered the open nest and were accepted by the parent robins.

A banded male Common Redstart (*Phoenicurus phoenicurus*) fed nestlings in a neighboring nest while also attending those in his own nest, who were about eight days younger. He brought food to his neighbors' nestlings about four times as often as their father did, unopposed by the latter. The helper's mate also tried to feed the neighbors' brood but was driven away by their mother. At a nest of the Black Redstart (*P. ochrurus*) a second adult male, in better plumage than the father, helped to feed the nestlings and to remove droppings. The parents were not hostile, but their efficiency in feeding their brood was impaired by this intrusion.

At several nests of the Ant Chat (*Myrmecocichla aethiops*) in northern Nigeria, three adults were in attendance. Four observers have found one extra male feeding the young at nests of the

Rufous Rockjumper (*Chaetops frenatus*), also in Africa, but it is not known how frequent this behavior is. With so many incidental observations of intraspecific nest helpers in this great family, it is not unreasonable to expect that sustained studies of permanently resident tropical species will reveal advanced cooperative breeding.

References: Armstrong 1947, 1955; Bent 1949; Brackbill 1943; A. J. Brown 1981; Carr and Goin 1965; Finley 1907; Forbush 1929; L. G. Grimes 1976; Ivor 1944b; Lack 1953; Laskey 1939; Mills 1931; Nice 1943; Nicholson 1930; Pinkowski 1975, 1978; Raney 1939; D. W. Snow 1958; Sutton and Parmelee 1954; Wynne-Edwards 1952; H. Young 1955.

37. *BABBLERS AND ALLIES*

Timaliidae

The 257 species of babblers and their relatives are distributed over much of the Old World, from Europe to the Philippines, Indonesia, Australia, and Africa, with a single species, the Wrentit (*Chamaea fasciata*), native to the New World along the west coast of North America. Ranging in size from 3.5 to 16 inches (9 to 41 centimeters), the members of this family are exceedingly diverse in plumage and habits. Their colors vary from the plain browns and grays of many babblers to the warm red and yellow of the Red-billed Leiothrix or Pekin Robin (*Leiothrix lutea*), a native of southern Asia introduced and now flourishing in the Hawaiian Islands. Their habits are so diverse that some have been compared to thrushes, others to wrens, thrashers, jays, titmice, pittas, and other unrelated families. Many are ground foragers, others arboreal. The family receives its name from the babbling chatter of some of its more sociable members; many are fine songsters. Their nests may be open cups or well-enclosed structures in trees, on rock ledges, or on the ground. Their two to seven eggs are variously colored.

A well-studied example of the cooperatively breeding babblers is the Gray-crowned Babbler (*Pomatostomus temporalis*) of eastern Australia, a 10-inch (25-centimeter), strong-legged bird with a fairly long, sharp, curved bill. Above, it is grayish brown, with

Gray-crowned Babbler

white-tipped tail feathers. It has a white eyebrow, dusky face, white throat and chest, and pale brown abdomen. The color of the iris changes gradually from the dark brown of the juvenile to the yellow of mature birds over two and a half years old, thereby revealing the age and status of individuals to their companions, as appears to be of importance to cooperative breeders. In open grassy woodland, these babblers live in stable groups of two to thirteen birds, including equal numbers of males and females. Almost wholly insectivorous, they forage on the ground and over trunks and branches of trees but rarely amid foliage. On the ground they overturn large objects by inserting their bills under them and pushing forward and upward. On trees and logs they pry off loose flakes of bark. They probe crevices and holes to the full length of their bills.

Brian King, who studied Gray-crowned Babblers for three years in southern Queensland, recorded many manifestations of their close companionship. Often one spontaneously offered food to a group member, of whatever age, who had not begged for it. The recipient might pass it to another bird, the latter to a third. In late afternoon, before retiring to rest, they dust-bathed close together, flicking dry earth or sand over themselves and their neighbors with their bills, while they fluffed out their plumage and fluttered their wings. Then they would fly up to a tree and diligently preen each other as well as themselves, sometimes three birds together attending to a fourth. Whether they arranged their feathers or drowsed on a perch, they rested in close contact. At night the whole group, except the primary female—who might be incubating eggs or brooding nestlings alone in a brood nest—crowded into a roofed dormitory nest, of which several were always available in each territory and kept in good repair by all group members. Unlike many other birds, these babblers apparently never slept in old brood nests.

Groups of babblers defended territories of 11 to 37 acres (4.5 to 15 hectares) or foraged over much larger home ranges. In each group, the single breeding male was dominant, the single breeding female next in the hierarchy, while the others ranked according to their ages, the youngest lowest. Aggressive encounters were rare. When a member of the breeding pair disappeared, it might be replaced either by the advancement of a nonbreeding adult or subadult of the same group or by an immigrant from another group.

An unusual activity of Gray-crowned Babblers was the huddle display, which began when the primary birds came together on a branch and rapidly repeated chuckling or barking notes. At this signal, other group members flew up to them and voiced the same notes while, crowded together, they postured with bodies held low, tails fully spread, wings partly extended and fluttering. Soon the mated pair stood upright amid the chuckling crowd and called antiphonally. If a huddle occurred while the breeding female was in her nest, one or more nonbreeding females joined the breeding male in antiphonal calling. These displays were given at all seasons and whenever a trespassing group or lone babbler was encountered in the home territory, as though to demonstrate, or to reinforce, the resident group's solidarity. In boundary disputes, each group huddled on an exposed perch in view of the other, sometimes continuing for as long as five minutes. Between repeated huddles, members of the opposing groups chased each other but rarely fought. While these confrontations continued, birds of one group might feed the begging young of the rival group, an occurrence which King saw eleven times.

When a babbler was caught by a predator or held in a bander's hand, it uttered distress calls which caused every babbler within hearing, of whatever group, to fly immediately to the caller and join in simultaneous distraction displays.

The Gray-crowned Babblers' nest is a bulky closed structure, entered through a short tunnel under an overhanging projection. Situated among the outer branches of a tree or shrub from 5 to 52 feet (1.5 to 16 meters) above the ground, it has an external diameter of 12 to 20 inches (30 to 51 centimeters) and contains a chamber as much as 8 inches (20 centimeters) in diameter. Within a frame of coarse twigs is an inner wall of fine twigs and grasses, lined with grass, feathers, thistledown, fibrous bark, and other soft materials. Like the dormitory nests, these brood nests are built by grown group members of both sexes. In such a nest the single breeding female present in most groups lays two or three eggs. An occasional larger set contains eggs differing in size, shape, and color, suggesting that they were laid by two or three females, members of as many pairs rather than consorts of a polygynous male.

During the long incubation period of twenty-one to twenty-five days, only the female sits, for intervals of about twenty-five minutes, rarely as long as an hour. At two nests, she was present 67 and 68 percent of the time. While in the nest she is visited by all group members, who bring nest materials or food with increasing frequency as the date of hatching approaches. During the twenty to twenty-two days that the young remain in the nest, all members of the group bring them food. At one nest, three nonbreeding helpers contributed more than the two parents together, although at all nests the father was the most active provider. After the young flew, they nearly always slept in the dormitory nest with their elders, all of whom responded to their pleas for nourishment. Now, in several groups, the mothers fed the young substantially more often than the fathers. Older group members fed both nestlings and fledglings more often than younger ones did.

Only one of King's nests yielded fledglings. From an analysis of the age composition of groups of Gray-crowned Babblers in southern Queensland, James Counsilman concluded that breeding success increased with the age of mated pairs and the number of their experienced helpers. The proportion of juveniles and fledglings was greater in larger groups. Groups with only two birds over one year old had an average number of younger companions of 0.4; with three experienced individuals, 1.3 young; with four, 1.8; and with six, 3.7.

Sixty-eight watches at eighteen nests by a team led by Jerram Brown demonstrated that the rate of feeding nestlings was independent of the number of attendants, being determined by the needs of the young and such environmental factors as temperature and rainfall. The parents benefited by having less work to do and saving energy as the number of their assistants increased. Similarly, the greater the number of helpers, the greater the saving of energy by each of them. Since the nestlings were fed at the same rate regardless of the number of feeders, the greater production of young by larger groups might be attributed to more nutritious meals delivered by attendants who had more time to select them or, more probably, to more effective defense of nests and fledglings. At two nests watchers found eight helpers.

Although less thoroughly studied, the other three Australian species of *Pomatostomus*, the White-browed Babbler (*P. superciliosus*), Chestnut-crowned Babbler (*P. ruficeps*), and Hall's Babbler (*P. halli*), have similar social and nesting habits and are known to be cooperative breeders.

In the lowland and foothill rain forests of New Guinea dwells a babbler in a habitat very different from that of its congeners in drier Australian woodlands. The 10-inch (25-centimeter) Rufous Babbler (*P. isidori*) lives in groups of two to ten individuals, who, like many other birds of rain forests, forage in mixed flocks of different species, three of which have similar rufous-brown plumage and mimic the babblers' notes. Other members of these flocks are birds of paradise and fantail flycatchers. Although Rufous Babblers occasionally ascend to the high forest canopy, they find their insect food chiefly amid low, dense vegetation and on the ground, where they rake up the litter and scratch open termite nests with their strong legs and feet.

The nest of the Rufous Babbler is very different from that of the Gray-crowned. One or, rarely, as much as 2 yards (1 to 2 meters) long, it is suspended from tendrils of the scrambling palm, *Calamus*, from 10 to 16 feet (3 to 5 meters) above the forest floor. Composed largely of palm and other fibers, it consists of a long column or stalk, an ample chamber with a narrow, hooded doorway in the side, and an appendage or "tail" hanging below the chamber. The general plan of these nests resembles that of the similarly elongated structures that Royal Flycatchers (*Onychorhynchus coronatus*) hang above woodland streams in tropical America, but the flycatcher lays its eggs in an open niche instead of a closed chamber. Harry Bell watched four Rufous Babblers building a nest and surmised that as many as eight were at work. Later, he saw at least three individuals feeding a young bird. These babblers, who have long been persecuted by Papuans, are very sensitive to human presence and difficult to study.

Over a period of five years, Anthony Gaston studied Common Babblers (*Turdoides caudatus*) in mixed scrub and dry deciduous woodland on the outskirts of New Delhi. Here and elsewhere on the Indian subcontinent, these babblers lived in groups of five to eighteen individuals, on home ranges that varied with the size of the group from about 7.5 to 15 acres (3 to 6 hectares). Adjoining groups seemed to avoid each other. When occasionally they came into conflict, they called loudly and chased one another but scarcely ever fought. Females stayed in their natal groups until they were six to nine months old, then dispersed. Males usually remained indefinitely in their natal groups, which might be considered male clans. Common Babblers foraged over the ground, seldom more than a few yards from some protecting bush, where when disturbed they promptly took refuge. Companions frequently preened each other, and at night they roosted side by side in a dense thorny bush instead of in their open nests.

Common Babblers at New Delhi bred in the spring, from March to May, and during the rainy season from July to December. In half the groups, only a single pair nested in a season, but in some large groups two pairs tried to breed, one always failing. Some males nested successively with two different females. Both members of a pair built the nest, in bushes so dense that observation was difficult, and only rarely was another individual seen to help. The three or four eggs in a nest were laid at the rate of one per day, evidently all by the same female. She alone incubated by night and performed slightly over 50 percent of the daytime incubation, with her mate contributing nearly 33 percent and one or more helpers sitting for 15 percent of the daytime. Nonbreeding females who approached a nest where the female parent was sitting were sometimes driven away by her. The average duration of the incubation period was thirteen and a quarter days, very much less than that of the Gray-crowned Babbler, which differs in many ways from the Common Babbler.

One to four nestlings were fed by the two parents alone or with one to nine helpers, who were more numerous when broods were larger and at some nests accounted for half or more than half of all feeding visits. The auxiliaries were mostly males; females were sometimes driven away from nestlings by their mother. The maximum feeding rate was ninety-three visits in four hours by nine helpers and two parents to three babbler nestlings and one cuckoo. Every bird who brought food also removed droppings, usually swallowing them on the spot. The more assistants the parents had, the less food they themselves brought. Parents with three or more helpers often continued to molt while their young were in the nest, but parents with only one or two auxiliaries arrested their molt, a difference which suggests that the assistants substantially reduced the burdens of parenthood. When, at the average age of 11.7 days, the fledglings left the nest, they were still unable to fly and usually spent several days perching separately in dense thorny bushes near the nest site before joining the group at the roost.

Of 37 nests of the Common Babbler, eggs hatched in 24 (65 percent), and 21 (57 percent) yielded fledglings. The average number of fledglings per group increased from 1.75 in groups of four or five, and 3.11 in groups of six or seven, to 4.25 in groups of eight or nine, then fell to 3.62 in groups of ten or more, suggesting that a pair with six or seven auxiliaries was the optimum size for breeding. Of breeding males, 88 percent survived from year to year, but the annual survival rate of breeding females and nonbreeding adults of both sexes was only 63 percent. Continuous residence in the natal group promoted longevity.

The Jungle Babbler (*T. striatus*), widely distributed in India, Pakistan, and Nepal, is, in the words of Thomas Gilliard, "as strange a bird as the traveler will ever encounter, awkward and gawky." The "seven sisters," as these birds are often called, live in stable groups of two to twenty individuals not only in forests, light deciduous jungles, and scrubby ravines but also in public and private gardens, shrubbery around villages, and university campuses. In compact parties, they hop over the ground rummaging in the litter, turning over leaves and digging vigorously in search of food, or they rise into trees to hunt for insects. They defend the territory in which they reside throughout the year. Young birds stay in the group in which they were reared for at least eighteen months, and those that disperse after this are chiefly females; males remain in their natal group for at least four years. Members of a group live in harmony; even at artificial feeding sites where food was concentrated, conflict was never seen, except among birds in their first year or between members of different groups. Parents with helpers build a low, open nest of ragged aspect but neatly lined, in which the female lays usually three to five bright blue eggs.

In dry deciduous woods and tropical thorn scrub near New Delhi, Gaston studied the social interactions of Jungle Babblers for three years. Like related birds, they perch in contact—a group member flying up to join several of its companions already resting in a compact row may alight on their backs and try to insert itself among them. Birds that build open nests rarely sleep in them, and Jungle Babblers are no exception. Amid the dense foliage or twigs of a tree or bush, most often from 10 to 15 feet (3 to 4.5 meters) above the ground, they line up side by side, all facing the same way, usually with the breeding pair pressed together in the center, a nonbreeding adult male on either end, and other group members between them. Fledglings from two to six weeks old often arrive last and squeeze into the middle of the row, where their companions help keep them warm. When between two and three months old, they roost in the outer third of the line but never at the extreme end, where a second-year or adult male continues to rest. The unsettled behavior and restless shifting around of babblers going to roost seem to be associated with the lack of discipline of younger birds.

Juveniles not infrequently engaged in two forms of vigorous play which Gaston called "rough and tumble" and "mad flight." The first was a mock fight between two or more birds, some of whom lay more or less passively on the ground while others rolled on top of them or pecked them deliberately but gently. Most of the four or more participants in these feigned conflicts were less than a year old; breeding adults were never seen to indulge in them. Gray-crowned Babblers and Large Gray Babblers (*T. malcolmi*) engage in similar rowdy play. In mad flights one or several birds flew rapidly and apparently aimlessly, wildly twisting and turning, among the branches of a tree. The participants in this play, too, were birds in their first year, never adults. Likewise, most instances of aggression between group members involved juveniles from one to three months old, who sometimes quarreled over food that they solicited from an older bird. As they grew older, Jungle Babblers learned to behave more decorously, during the day and as they went to roost in the evening.

With interesting results, Gaston took great pains to learn the relative frequency of participation in various activities by the several members of a group. While clumped together on a perch, babblers often preened their companions. Older individuals preened younger ones more often than the reverse; and in each group one of the two breeding adults, slightly more often the breeding male, was the most active preener. Birds in their first year solicited grooming more than any others, and by far the greater part of their solicitation was directed toward the breeding adults, who were usually their parents.

While a group of Jungle Babblers foraged over the ground, one of its members remained perching several yards up, keeping watch. In winter a sentinel was on duty most of the time. When, after five or ten minutes, it descended to resume foraging, another babbler flew up to replace it; or the watcher might remain at its post until the relieving bird alighted near it. In five out of six cases, the individual who served as sentinel was either the male or the female of the breeding pair, who together performed about 50 percent of this duty, while the rest was shared by nonbreeding adults and birds in their second winter; those less than a year old rarely undertook to watch. Apparently, the amount of time an individual spent as sentinel was related to its competence in foraging: the less time it needed to find food, the more it had for watching.

When a group of foraging Jungle Babblers moved onward, a breeding adult, more often the male than the female, usually led; other adults followed; and the youngest members came last. From these painstaking observations emerges the picture of a social group in which order is maintained by true leadership, competence, and service rather than by aggression and bullying. A peck order or clear-cut approach-retreat behavior is not evident among these birds. The highest-ranking members, the breeding pair, are the most active preeners, the most constant sentinels, the most frequent leaders. Preening of companions, which some ornithologists have interpreted as "redirected aggression," is among these babblers a friendly act which, like clumped perching and roosting, helps cement the bonds between group members. Although subordinate males might be expected to compete keenly for the post left vacant by the loss of the group's single breeding male, succession to this rank is effected in an orderly manner, without overt aggression.

As among other animals, adolescent Jungle Babblers are sometimes undisciplined, chasing adults for food, pecking each other rudely, and disrupting the roost. During their first winter their conduct improves, until by the following summer they behave like adults, although they still play minor roles in group activities. These "awkward and gawky" babblers have achieved an admirable society, the product of a long evolution. To adjust themselves to the complex personal relations of such a society, individuals seem to require something very like intelligence, discrimination, and tact. And this social integration promotes the survival of its members, which is so high that only 25 percent of mature birds ever breed.

Another Indian species that nests cooperatively is the Large Gray Babbler. In Africa, cooperative breeding is known to occur in no less than four species of *Turdoides*—the Brown Babbler (*T. plebejus*), Arrow-marked Babbler (*T. jardineii*), Black-lored Babbler (*T. melanops*), and Black-cap Babbler (*T. reinwardii*). In all these Indian and African babblers, one or more helpers have been seen feeding nestlings or fledglings. Helping with nest construction, preening of companions, and roosting in contact have been recorded for some of them and probably occur in all.

In the deserts of Israel, color-banded populations of Arabian Babblers (*T. squamiceps*) were studied for several years by Amotz Zahavi. In groups of two to fifteen individuals, these birds lived throughout the year on defended territories. Young birds, distinguished by the color of their bills, remained in their parental group for at least a year, through the breeding season following that in which they hatched. On reaching maturity, females tended to emigrate alone or in parties of two, three, or four. In eight groups, all females were replaced by aliens, but four groups were joined by female strangers without displacing all the original members of this sex. These transfers from group to group helped avoid inbreeding. In contrast to the females, most males remained attached to their natal groups.

Each group of Arabian Babblers built only one nest, and in nearly all of them three or four eggs were laid on successive days by one apparently monogamous female. A few nests contained eight or nine eggs, evidently the product of two or more females. Losses of eggs from these large sets suggested conflicts between the females who laid them. All group members fed and guarded the nestlings and fledglings. It was surprising to see young birds, who a few months earlier, as nestlings and fledglings, had fought over food and even attacked older group members, peaceably cooperating in the care of nestlings, as though there were no longer motives for competition among the nonbreeding group members. The average number of yearling birds in large groups was no greater than that in small groups, which led Zahavi to ask whether helpers really help—a question that was answered in the affirmative by Jerram Brown. Zahavi regarded a group of Arabian Babblers as essentially a coalition of birds for territorial defense.

The Yellow-eyed Babbler (*Chrysomma sinensis*), a small insectivorous bird that dwells in scrub, tall grass, and thickets at the forest edge from Pakistan to China, differs in its social arrangements from the foregoing babblers. In the cool winter weather, foraging groups of four to twenty birds move rapidly through scrub and lantana thickets; at night they roost packed together side by side on a branch, all facing the same way. They preen their companions. In February, these groups disintegrate into pairs, which sing with loud warbles and display against other pairs in seeming territorial defense. With the advent of the rainy season in late June or July, each pair builds its own nest, with the result that several may be found within the group's foraging range, instead of the single active nest usual among babblers of the genus *Turdoides*. The parents are sometimes accompanied by one or two other birds who apparently help feed the young. The breeding system of Yellow-eyed Babblers resembles that of Long-tailed Tits (*Aegithalos caudatus*), as described in chapter 29.

REFERENCES: Andrews and Naik 1965; H. L. Bell 1982; J. L. Brown 1975; Brown and Balda 1977; J. L. Brown et al. 1978; Counsilman 1977a, 1977b; Gaston 1977, 1978a, 1978b, 1978c; Gilliard 1958; L. G. Grimes 1976; Hutson 1947; King 1980; Vernon 1976; Zahavi 1974.

38.

MOCKINGBIRDS, THRASHERS, AND CATBIRDS

Mimidae

The thirty species in this family are confined to the western hemisphere, from southern Canada to southern South America and from the Lesser Antilles to the Galápagos Archipelago. From 8 to 12 inches (20 to 30 centimeters) long, they are slender, long-tailed birds with short to long and downcurved bills. Except two species in Mexico and northern Central America that are extensively blue, members of this family lack spectral colors but are clad in grays and browns, often with paler or white underparts that may be plain, streaked, or spotted. Inhabitants of thickets, forest edges, brushlands, and deserts, they forage for insects and other invertebrates on the ground, tossing leaves aside or digging with their curved bills, and they augment their diet with berries and larger fruits. They sing generously with strong, melodious voices; some are famous mimics. In often bulky, bowl-shaped nests of twigs lined with finer materials, situated in bushes and trees at no great height, sometimes on the ground, they lay from two to five, rarely six, whitish, buffy, greenish blue, or blue eggs that may be immaculate or spotted. Usually only the female incubates, but both sexes feed the young. Frequently aggressive, mockingbirds and thrashers valiantly defend their nests.

For four months on Isla Genovesa (Tower Island) in the Galápagos Archipelago, the Grants studied Galápagos Mockingbirds (*Nesomimus parvulus*), which after the breeding season live in territory-defending bands of up to forty individuals. At seven of twenty-nine nests, the young were fed by helpers, two of whom were juveniles of the first brood assisting their parents with the second brood. Two other banded helpers, in adult plumage, were apparently males. One of them aided a banded pair to rear two broods. Another assisted one pair with its first brood and a different pair with its second brood. Male parents and their adult auxiliaries delivered food either directly to the nestlings or to their brooding mother, who usually passed it to the young but occasionally ate it herself. At second-brood nests the sibling helpers always fed the nestlings directly. Some helpers fed the nestlings as frequently as their father, others much less, but the mother was always the chief provider. As in other species with helpers, the auxiliaries did not consistently increase the frequency of feeding visits, which, when food is adequate, is determined primarily by the needs of the young rather than by the number of attendants.

Helpers increased the productivity of nests. Of twenty-two nestlings attended by them, nineteen or 86 percent fledged, whereas of thirty-eight nestlings fed by the parents alone, twenty-six or 68 percent survived to leave the nest when eleven to seventeen days old. On Indefatigable Island, Jeremy Hatch had earlier watched a recently fledged Galápagos Mockingbird beg unsuccess-

Galápagos Mockingbird

fully, then sing briefly and feed nestlings of its parents' next brood.

At two of fifty-nine nests of Chalk-browed Mockingbirds (*Mimus saturninus*) watched in Argentina by Rosendo Fraga, helpers were present. At one of these nests, the helper was a yearling female offspring of a neighboring pair, at whose nest it was banded. At the other nest, attended by three mockingbirds, the helper was of unknown age and parentage. Young Chalk-browed Mockingbirds of the season's last brood remain with their parents throughout the ensuing nonbreeding season. In Venezuela, I found juvenile Tropical Mockingbirds (*M. gilvus*) closely associated with their parents while they raised later broods, sleeping quite near the nest and helping defend it. Future studies will probably reveal that nest helpers are widespread among mockingbirds.

In the migratory Gray Catbird (*Dumetella carolinensis*), widespread in the United States and southern Canada, intraspecific helpers are apparently unknown, but interspecific helpers have been reported. A catbird fed and mothered a brood of orphaned Northern Cardinals (*Cardinalis cardinalis*). Another catbird fed a half-grown Northern Flicker (*Colaptes auratus*) that had been dislodged from its nest and separated from its parents in a severe storm. A female catbird who was beginning to incubate fed nestling Northern House-Wrens (*Troglodytes aedon*) in a neighboring box. At first she brought food frequently, about once every ten minutes, but after incubation was well advanced she fed the nestlings at much longer intervals, carrying food to them as she returned to her eggs after a recess. The parent wrens did not oppose their assistant, but after some days the helpful catbird and her mate sometimes chased the wrens when they met them.

References: Bent 1948; Fraga 1979; Grant and Grant 1979; Hatch 1966; Kinnaird and Grant 1982; Nolan and Schneider 1962; Skutch 1968.

39.

ACCENTORS

Prunellidae

The twelve species of accentors, all in the genus *Prunella*, are widely distributed in the Old World from the British Isles to Japan and Mediterranean Africa, with outposts in Taiwan and southwestern Arabia. One species, the Siberian Accentor (*P. montanella*), is casual in Alaska in autumn. From 5 to 7 inches (13 to 18 centimeters) in length, they are sparrowlike birds with streaked, rufous or brownish gray upper plumage and grayish underparts, often with rufous markings. The sexes are alike. One species of these hardy birds, the Himalayan Accentor (*P. himalayana*), nests far above timberline, sometimes as high as 17,000 feet (5,180 meters); others endure the harsh winters of their far northern homelands. With bills more slender and pointed than those of sparrows (to which they are not closely related), accentors hop over the ground in bushy places, seeking principally insects in summer, seeds and berries in winter. They sing persistently in rather subdued voices. Their nests are usually open cups placed low in a bush or in a crevice among rocks, according to the species. The three to five bluish or bright blue eggs are incubated by the female only, but both parents feed the young.

In the mountains of Poland, Andrzej Dyrcz found a single adult helper at two of the three nests of the Alpine Accentor (*P. collaris*) that he watched. One of these auxiliaries was a male who had hatched in the preceding year in a nest near that at which he assisted; he may have been a son of the breeding pair. These two helpers contributed much less food than the parents brought.

On the grounds of a school in Edinburgh, Scotland, Michael Birkhead studied for two years a small, color-banded population of Dunnocks (*P. modularis*, often miscalled Hedge-Sparrows). In winter these birds occupied defended territories singly, in pairs, or in groups consisting of two males, or two females, or two males and one female, or two males and two females. In these groups dominance hierarchies developed, with the birds graded in the order of their weights, the heaviest at the top. Since females were lighter than males, they consistently ranked lower and were often supplanted at feeding sites, which may have resulted in emaciation or starvation in severe weather. This, coupled with the higher mortality often suffered by the incubating sex, probably explains why during the breeding season the population consisted of twenty-one males with only fifteen females.

Breeding groups consisted of five unaided pairs, four pairs each with one male helper, and a polygynous male with two females and one male helper. At three of ten nests, a male auxiliary brought food during an observation period of three hours. In contrast to the usual results of cooperative breeding, and in spite of the fact that groups of three had significantly larger territories than pairs, nests with helpers were much less successful than those of unaided pairs, producing only 1.75 fledglings per group, while the pairs produced 4.6. Except at the nest with the most diligent helper, nestlings nourished only by the two parents gained weight more rapidly than those with three attendants. The poor results of nests with helpers appeared to be a consequence of the lack of harmony between the male parent and his assistant, who, when they met at the nest, often dropped food intended for the young while they displayed to each other. Moreover, members of pairs appeared to bring much more food than members of trios on each visit.

Dunnock

Mutual preening, resting in contact, exchange of food between adults, and other expressions of intimacy frequent among successful cooperative breeders appeared not to occur among these Dunnocks. In the two years during which Birkhead watched them daily, the only occasions when he ever saw physical contact, except briefly in coition, were when they fought. Members of a group did not cooperate in territorial defense. Widespread in the tropics and subtropics, cooperative breeding is unusual at 56 degrees north, the latitude of Edinburgh. The poor success of nests with auxiliaries in this study helps us understand why cooperative breeding is so rare in lands where severe winters intensify competition for food in the relatively few species that remain on their territories instead of wandering or migrating to warmer regions.

References: Birkhead 1981; Dyrcz 1977.

40.

WOOD-SWALLOWS

Artamidae

The Australasian region, comprising Australia, New Guinea, New Zealand, and neighboring smaller islands, is the scene of some of the most fascinating developments in the avian world, including the extravagantly ornate plumages of the birds of paradise, the decorated gardens of bower birds, and the curious incubators of megapodes. It also contains a surprising number of cooperative breeders, among which the wood-swallows are outstanding for their multifarious mutual helpfulness as well as their flight. Thomas Gilliard, who knew them in New Guinea, called them "the finest of passerine fliers, and excepting the ravens the only songbirds known to soar."

Wood-swallows are found from India to the Philippines and the Fiji Islands and southward over Australia, where six of the ten known species occur. From 5 to 8 inches (13 to 20 centimeters) long, they have stout bodies, pointed wings that reach almost to the ends of their short tails, short necks, stout, moderately long, pointed bills, and short but strong legs and feet. Their soft plumage is variously brown, gray, black, and white. One of the largest and most richly colored species, the White-browed Wood-swallow (*Artamus superciliosus*), is sooty black above and rufous-brown below. The sexes are similar. Another feature in which wood-swallows differ from all other songbirds

White-breasted Wood-swallows

is their powder-down feathers, the tips of which disintegrate into a fine powder for dressing the plumage. The family appears not to be closely related to any other, certainly not to the true swallows.

Wood-swallows subsist wholly upon insects, which they capture on flycatcherlike sallies from exposed branches or overhead wires, while they glide on set wings, or while the larger species soar high in the air on thermal updrafts. Rarely they gather food from the ground, sometimes while following the plow. They utter harsh cries, swallowlike twittering songs, and imitations of the calls and songs of other birds. Both mated and bachelor males sing throughout the year. They build loose, shallow cups of fine grasses, rootlets, feathers, and similar materials, high in a tree, low in a shrub, at the bases of palm fronds, or, as in the Little Wood-swallow (*A. minor*) of Australia, in crevices in cliffs or on ledges in caves. Their two to four white or cream-colored eggs are spotted and blotched with brown or lavender, most heavily in a wreath around the thicker end.

Klaus Immelmann, who for ten months studied three species of wood-swallows in Australia, was impressed by their highly social nature. Unlike true swallows, who when resting in a row perch a few inches apart, maintaining their individual distance, wood-swallows line up in close contact with their companions. If one in the middle flies away, its neighbors close the gap. A new arrival often alights on the backs of those already present and tries to insert itself among them. While perching in contact, they preen one another, not confining this attention to members of their immediate family but billing whoever is nearest. An individual may be groomed simultaneously by neighbors on both sides. Siblings preen each other on the day after they leave the nest. Wood-swallows forage together, sometimes two or three pursuing the same insect; they were never seen quarreling over food. They sun themselves and bathe in intimate groups which may contain more than one species. Only a few days after they fledge, young wood-swallows join their elders in spirited attacks upon predatory birds from crows to eagles, with loud alarm calls and menacing dives driving them from the vicinity of their nests.

At night one member of a pair roosts beside its incubating partner, but throughout the year members of a group not so engaged sleep in closest contact, either lined up on a branch, massed in a compact ball, or hanging from a branch or the bark of a tree, in a manner not well understood. Sometimes such clusters of sleeping birds hang in a hollow tree trunk. From ten to fifty individuals often clump together in this fashion, occasionally as many as one hundred. A latecomer approaching a roosting group is greeted by an outburst of cries. The Black-faced Wood-swallow (*A. cinereus*) often flies in carrying a twig, which it drops. By huddling together at night, wood-swallows conserve heat and energy. If disturbed by a nocturnal predator, the mass of sleeping birds might explode in all directions with a sudden burst of wingbeats that would alarm the hungry animal.

It would be surprising if birds that live so intimately did not cooperate in reproduction. Although they do not defend nesting territories, wood-swallows are not colonial. However, two pairs of Black-faced Wood-swallows may nest in neighboring bushes or in the same tree only 6 feet (2 meters) apart. Keeping close company, a male and female build their open nest with materials gathered from the ground, including roots that they strenuously pull up. Both sexes incubate, generally for very short intervals; while one covers the eggs, the other often rests beside it. The male White-browed Wood-swallow usually incubates at night. After about twelve days of incubation, the nestlings hatch and are fed not only by both parents but also by neighboring pairs. As many as six Black-faced Wood-swallows may come and go together, by turns nourishing the young in their respective nests. Juvenile White-breasted Wood-swallows (*A. leucorhynchus*) help their parents feed a later brood. At twelve days of age the fledglings leave the nest, and a day or two later they join the young of other families in a tree, where all the parents feed them in common, those of other pairs along with their own.

This many-sided helpfulness does not end here. White-breasted Wood-swallows continue to feed their young for weeks after they become self-supporting. Independent juveniles give food to unrelated young as well as to their siblings. Adults feed other adults, not confining this attention to their mates. In the Kimberleys, Immelmann found an obviously ailing Black-faced Wood-swallow who was apparently kept alive only by the food that several companions brought it throughout the day. When one wood-swallow alights beside another, they greet each other with postures and movements not unlike those of fledglings begging for food. A newcomer arriving with food in its bill may posture in this manner in front of another who has none. This behavior rarely leads to the transfer of food but may stimulate the settled bird to preen the new arrival, or the latter may groom the former. Feeding and being fed, mutual preening, sunning and bathing together, perching and sleeping in contact all help bind wood-swallows into societies of closely cooperating individuals.

The widespread White-breasted Wood-swallow extends from Australia through the islands of the southwestern Pacific as far as Fiji, where the race *A. leucorhynchus mentalis* occurs. Here Fergus Clunie studied a group of four and occasionally five or six individuals who for several years were permanently resident, roosting and nesting in a large rain tree (*Samanea saman*) overhanging a busy intersection just outside the commercial center of the capital city, Suva. These urban wood-swallows mostly ignored people but menaced and drove from their tree and its vicinity both predatory and innocuous birds. They also harried dogs passing beneath the rain tree and cats within 50 yards (46 meters) of it. One of the latter retaliated, pulling many flight feathers from a wood-swallow's wing. After catching a large insect in the air, these wood-swallows often transferred it to their feet while they flew to a perch—raptorlike behavior most unusual in passerine birds, who nearly always use only their bills for carrying. Or they might repeatedly drop and recatch the insect.

Like their relatives in Australia and New Guinea, these wood-swallows perched by day and slept by night pressed closely together, repeatedly preened each other, and frequently gave food to their adult companions. Even while attending young in the nest, they fed and kept alive an injured member of their group incapable of foraging for itself, much as the Black-faced Wood-swallows that Immelmann watched in the Kimberleys cared for a sick individual. Although Australian wood-swallows apparently regularly breed in monogamous pairs, these Fijian wood-swallows nested in a promiscuous quartet of two males and two females. Together the four chose the nest site in the rain tree, testing promising forks by clustering together in them or circulating through the tree in a follow-the-leader procession, while one sat as though incubating in a potential site. On the day after this ceremonial site choosing, the four birds started to build the first of a series of preliminary nests, each bringing strands of dry grass, sometimes in their feet, laying them in the fork, and weaving them together with their bills. Building continued during four months, from May to September, before eggs were laid.

All four birds took turns at incubation, replacing each other frequently; while sitting they were fed by others. Sometimes the incubating bird was reluctant to make way for another. When this occurred, the newcomer tried to push the stubborn one from the nest, climbed on top of it, or sat beside it until one or the other left. After three nestlings hatched, they were fed at intervals of a few minutes by the four adults, who sometimes queued up to deliver their offerings. Apparently the Fijian race of the White-breasted Wood-swallow often breeds in similar quartets, for on outlying islands of the archipelago two other groups of four built nests in holes in cliffs. Whether both females lay eggs in the same nest and whether one of the males is dominant over the other are questions that remain to be answered.

To the four species of Australian wood-swallows, the White-breasted, Little, Black-faced, and Dusky (*A. cyanopterus*), now known to breed cooperatively can be added the Papuan Wood-swallow (*A. maximus*) of the mountains of New Guinea. At a nest of this sooty-backed, white-breasted bird 50 feet (15 meters) up in a fork of a dead tree, the photographer Loke Wan Tho watched four or five adults feed three well-feathered young.

REFERENCES: Clunie 1976; Gilliard 1958; Immelmann 1966; Rowley 1976.

41.

SHRIKES

Laniidae

Only two species of shrikes—the bold grayish, black, and white "butcher-birds" that impale insects and small birds and mammals on thorns and barbed-wire fences—inhabit North America, from northern Canada to southern Mexico; none is found in Central or South America. The remaining sixty-eight species of the family are widely spread over the Old World from Eurasia to the Philippines and New Guinea and southward over Africa. Ranging in size from about 6 to 14 inches (15 to 36 centimeters), they have stout, sharply hooked, notched bills and strong feet with sharp claws. In contrast to the subdued colors of the true shrikes, the numerous bush-shrikes of Africa are often bright with red, yellow, or green. Also unlike the true shrikes, some are melodious songsters, notable for their duetting or antiphonal singing. Not all shrikes prey upon other birds; many are largely insectivorous. After building, in trees or shrubs, open nests that may be bulky or small, they lay from two to eight eggs, pale pink, brownish, or greenish, sometimes bright green, marked with shades of brown. Often incubation is chiefly by the female, fed by her mate, but in some species both sexes warm the eggs.

Among the more social members of the family is the medium-sized, dusky Yellow-billed Shrike (*Corvinella corvina*), which inhabits savanna wood-

Yellow-billed Shrike

lands of Africa north of the equator, from Gambia to east of the Sudan. For six years, it was intensively studied by Llewellyn Grimes on the extensive park-like campus of the University of Ghana at Legon. There the birds lived in cooperating groups of five to, exceptionally, twenty-five, permanently resident on territories—defended by all adult group members—that ranged from 26 to 63 acres (10.5 to 25.5 hectares), with no correlation between the area of the territory and the number of birds in the group. A single group might contain, in different years, more males than females or the reverse; in the population as a whole, the sexes were present in approximately equal numbers. No dominance hierarchy was evident but, as in other cooperative breeders, it may have been maintained by such subtle interactions between group members that it escaped detection.

On their territories, Yellow-billed Shrikes caught wingless ants, grasshoppers, and small lizards either by swooping down from a perch or pouncing after a low, short flight. When visibility was poor, they foraged over the ground, busily rummaging in the litter. Green mantids, caterpillars, spiders, slugs, and earthworms augmented their diet. When winged termites filled the air, shrikes captured them in flight. At the day's end, they flew singly or by twos to the tree where the whole group roosted.

Yellow-billed Shrikes laid eggs in every month except October, but chiefly toward the end of the dry season in January and February and into the rainy season from March to July. The nest, an open structure of slender twigs lined with finer, more pliable tendrils, was placed in the fork of a tree or astride several branches, usually between 10 and 20 feet (3 to 6 meters) above the ground. Both the breeding female and her presumed mate shaped the nest with twigs brought by other group members of both sexes, who also fed the breeding female. Although some groups built only a single nest before eggs were laid, in others two or three females, competing for breeding status, started or completed nests. While building, they called and were visited by other group members, who flew from one to another. Laying was postponed until, in some obscure manner, one of these hens had established herself as the only breeding female in the group. Then, on consecutive days, she deposited from three to five eggs, usually four. In a single group, two females never bred simultaneously.

While sitting on her eggs, the female shrike repeated begging cries, similar to those of nestlings, that attracted her mate and all other members of the group with food for her. Occasionally a male other than her mate sat in the nest, begged, and was fed, but he was soon displaced, with much scolding, by the breeding female. After an incubation period of sixteen to eighteen days, the young hatched. For the next eight or nine days their mother brooded them closely, received all the food brought by group members, and distributed it to the nestlings. In her absences, her helpers were generally reluctant to feed the young directly before they were about a week old. Occasionally, however, nonbreeding females did feed them while their mother was away, and they also begged successfully for food brought by other shrikes. The mother also brought food both during and after the period when she brooded the nestlings by day, but with so many assistants, she stopped feeding when they were about twelve days old. Food was brought at the average rate of one or two meals per nestling per hour when they were newly hatched, increasing to twelve per hour when they were two weeks old, but the number of meals in single hours fluctuated widely.

When nineteen days old and scarcely able to fly, the young shrikes fluttered helplessly to the ground. They could scramble and climb to suitable cover, guided when necessary by solicitous group members. All attended the young at one time or another, but individual shrikes did not take charge of single fledglings—a practice which many birds find efficient. In their seventh week after hatching, the young shrikes became independent; those who continued to beg beyond this age were mostly ignored by the adults. By their tenth week they were well-integrated members of their group, gave alarm calls, roosted with the adults, and participated in social displays. When about fourteen weeks old, these young birds fed fledglings, and at the age of twenty-four weeks they brought food to nestlings. Those who survived tended to remain in their natal group for a long while. The few that were known to migrate to other groups, usually at no great distance, were over two years old and occasionally as much as five years of age. As in other cooperatively breeding birds, they sometimes changed groups in doublets or trios of the same sex. At the end of Grimes' six-year study, he still did not know the age when Yellow-billed Shrikes begin to breed. Two females in their sixth year had not become breeders.

No breeding group at Legon had fewer than five members, and breeding pairs without helpers were not known anywhere in the Yellow-billed Shrike's range. Apparently, auxiliaries are indispensable for successful nesting, as in the White-winged Chough (*Corcorax melanorhamphus*). Helpers who participate in territorial defense, guarding the nest, and feeding the incubating female and her young reduce the burden of the parents and enable them to breed more frequently; at Legon they laid from one to five clutches and raised two or three broods a year. Always the clutch following a successful nesting was laid while group members other than the single breeding female cared for the fledglings of the earlier brood. Despite the many guardians and attendants of the nests, only 25 percent of the total number of eggs laid yielded fledglings. Of sixty-six banded nestlings that left their nests, twenty-nine (44 percent) were alive a year later.

The related Magpie-Shrike (*Corvinella melanoleuca*) also breeds cooperatively, but details are not available.

References: L. G. Grimes 1976, 1980.

42.

HELMET-SHRIKES

Prionopidae

Although sometimes classified with the true shrikes in the family Laniidae, the nine species of helmet-shrikes are so different that they are now more often placed in a separate family. Highly social birds from 7.5 to 10 inches (19 to 25 centimeters) long, they wander through woodlands and savannas of Africa south of the Sahara in small parties, hunting the insects that are their chief food. Several pairs often nest close together. One of the best-known species is the Straight-crested, or Common, Helmet-Shrike (*Prionops plumata*), an 8-inch (20-centimeter) bird widespread on the more southerly parts of the continent. Largely blackish above, it has white underparts, a white collar, and a white-tipped tail. Its grayish head bears whitish plumes that project forward over the nostrils and curve upward as a bushy crest above the crown. Its pale eye is circled by a prominent yellow wattle.

Carl Vernon found Straight-crested Helmet-Shrikes living in groups of two to twenty-two birds, usually six to eight, who forage, roost, and breed together, keeping in contact by uttering varied notes and loudly snapping their bills. They return each year to the territory that they occupied in the preceding breeding season, where all group members help build a neat, shallow, cup-shaped nest in a tree. The female of the single dominant breeding pair lays three or, more often, four eggs, rarely supplemented by one or more that a subordinate female adds to the nest. During the incubation period of seventeen days, all group members participate rather equally in warming the eggs, and during the further seventeen days of the nestling period, all bring food to the young. Often several arrive at the nest together, and the last to deliver its contribution remains to guard it. Bold in defense of their eggs and young, helmet-shrikes may attack large animals or even people who approach them too closely.

The young remain in their group until the approach of the following breeding season, when single individuals or several birds of the same sex, either males or females, go in search of similar parties of the opposite sex to form new groups, which consist of a set of male siblings joined to a set of female siblings. If some of these emigrants return to their natal group, they are freely accepted. The breeding male and female of a new group appear to be the oldest members of their respective sexes. If one dies, its place is inherited by its sibling next in age.

At least three other species of helmet-shrikes breed cooperatively.

REFERENCE: C. J. Vernon, *in litt.*

43.

STARLINGS

Sturnidae

The 110 species of the starling family are all native to the Old World, where they are most abundant in India, Indonesia, and Africa south of the Sahara. Through misguided human agency, one species, the Common Starling (*Sturnus vulgaris*), has spread widely over the Earth, including North America, where it has become excessively abundant and troublesome, although Central and South America have so far escaped this plague. Likewise with human help, the Common Myna (*Acridotheres tristis*) has become established on continents and islands far from its original home in Asia.

Members of this family range in length from about 6 to 16 inches (15 to 41 centimeters), including species with very long tails. Many are handsome birds, with dark glossy plumage, often with large areas of white, yellow, or purple. The mynas, especially the Hill Myna (*Gracula religiosa*) of India and Malaya, are famous for their ability to imitate sounds, including human speech. Starlings eat both fruits and insects, the latter often gathered from the ground, over which they walk instead of hopping like many passerine birds. Often they gather in huge flocks, such as those of the Rosy Pastor (*Sturnus roseus*) of southeastern Europe and southwestern Asia and the Wattled Starling (*Creatophora cinerea*) of Africa, both of which follow hordes of locusts and nest opportunistically wherever they find these grasshoppers so injurious to crops reproducing abundantly. Many starlings make crude nests in holes in trees or whatever crevices and crannies are available, while others build covered nests with a side entrance in trees and shrubs. They lay three to seven white or tinted, immaculate or variously marked eggs, usually incubated by the female alone. Both parents feed the young.

An example of the numerous African species that breed cooperatively is the Chestnut-bellied Starling (*Spreo pulcher*), which Roger Wilkinson studied for three years on and around the extensive campus of Bayero University at Kano, Nigeria. Here, in savanna country, the starling nested amid the introduced trees and gardens around the university buildings and on farmlands surrounding the campus. The birds live in social groups of nine to thirty-five individuals, each of which may contain from two to six breeding pairs, nonbreeding adults of both sexes, and juveniles. Alike in plumage, adult males and females can be distinguished only by behavior. Fledglings lack the glossy chestnut breast of adults and have dark eyes and yellow bills instead of the cream or white eyes and black bills of adults. Although when about six months old they resemble the adults in plumage and in bill and eye color, males do not breed until about two or sometimes three years of age. At Kano, males outnumbered females by approximately 1.3 to 1. They tended to remain and breed in their natal group, whereas females more often migrated to neighboring groups, as in other cooperatively breeding birds.

In this region of two annual wet and dry seasons, the starlings have two breeding seasons each year, one before and one after the main rains, which fall from July to September. A peculiarity of this starling is that in a single group two to six pairs nest simultaneously. A breeding male and female share rather equally the work of building, chiefly with dry grasses, a dome-shaped nest with a side entrance, which they place in a low, thorny tree, high in a clump of mistletoe, or on abandoned nests of sticks made by the White-billed Buffalo Weaver (*Bubalornis albirostris*). The female, incubating alone, is rarely fed on the nest by her mate.

After the nestlings hatch, helpers begin to feed them and to remove their droppings. Although they rarely participate in building the nest, occasionally these assistants bring grasses or feathers to it while it holds young. As the nestlings grow older, the number of these attendants increases, until before fledging as many as fourteen are sometimes present, although seven is more usual, and a few nests lack helpers. Juvenile starlings begin to feed nestlings as early as six weeks after they leave the nest, but they tend to restrict their visits to later nests of their own parents. Yearlings spread their attentions more widely. The most active assistants are birds in their second year, who help, simultaneously or successively, at nearly all the nests of different breeding pairs available to them in their group. Adult nonbreeding males distribute their aid among the different nests about as widely. Breeding males helped at nearly half of the nests at Kano. Breeding females tended to feed only at their own nests but occasionally assisted at their neighbors' nests. Nonbreeding females, who often migrated to a different group, distributed food among the nests only slightly more widely than the breeding females did. Rarely a starling took food to a nest of a group adjacent to her own, but she might have done so more often if the resident birds had not repulsed her. After the young left the nest, helpers as well as parents continued to attend them.

Because so few nests lacked helpers, it was not possible to compare the success of those with and without them. Nests with seven to fifteen attendants yielded significantly more fledglings than nests with only two to six attendants. However, this could not with confidence be attributed to the number of helpers, because the former nests contained more eggs than the latter. Nevertheless, the advantage of having helpers was suggested by the fact that nests with more than seven attendants fledged 82 percent of the nestlings that hatched, whereas those with six or fewer attendants fledged only 63 percent. Although the Chestnut-bellied Starling's breeding system, with active nests of several females attended simultaneously or successively by a number of group members, is less usual among cooperative breeders than a single reproductive pair whose nest receives all the attention of group members, it nevertheless appears to be a highly successful system.

The related Superb Starling (*Spreo superba*) of East Africa follows a similar system, with groups of up to twenty individuals, including several breeding females, who lay and incubate their eggs in separate nests. From mid October to the end of the following June, the six females of one group nested at least twenty-two times. One of them laid six sets of eggs, of which three produced fledglings. At the nine successful nests of this clan, the average number of attendants feeding nestlings was 12.2. Only a month after they leave the nest, juveniles help at other nests.

The long-tailed, slender, iridescent green and violet Golden-breasted Starling (*Cosmopsarus regius*), also of East Africa, lives in groups of from three to twelve individuals and nests during the rainy season in holes in trees that were probably carved by woodpeckers. Apparently, only the female incubates, but

Superb Starlings, adult (left) and immature (right)

she may be accompanied back to her nest by up to eight individuals. At one nest, at least three of these escorts passed nest materials to her and at least two fed her. At a second nest, five Golden-breasted Starlings fed an undetermined number of nestlings, and at two additional nests five and six birds were active. Sometimes the attendants fed the nestlings directly, but often they surrendered the food to the soliciting female, who in turn passed it to the young.

About seven pairs of Pied Starlings (*Spreo bicolor*) nested under the eaves of a stone-walled shed near Grahamstown in South Africa. Only a single pair entered a nest until the nestlings hatched, when they were fed by nine attendants, including the parents, a pair of adults who were building their own nest nearby, and five subadults, one of whom had helped at the nest of another color-banded pair earlier in the season. On the same South African farm was a nest of Cape Glossy Starlings (*Lamprotornis nitens*) in an oak tree, at which three of the five adults that were present fed the nestlings. Cooperative feeding of young has also been recorded of Fischer's Starling (*Spreo fischeri*).

The two species of oxpeckers are brown, stout-billed, short-legged, 7-inch (18-centimeter) starlings that with sharp claws cling woodpeckerlike to the skins of a number of grazing African animals, from domestic cattle and antelope to rhinoceroses and giraffes, pulling off and eating the ticks that infest them. Flies, often caught on the wing, supplement this principal fare. Since the mammals that supply the oxpeckers' food are constantly moving about, to defend territories would be useless, and up to a dozen birds may forage simultaneously on the same animal. They roost in large companies in trees, reeds, cliffs, or buildings. Oxpeckers nest in ready-made holes high in trees or in crannies in stone walls, buildings, or embankments, where they lay from two to five eggs, tinted with blue or pink and variously spotted. Four adults sometimes attend a single brood of the Yellow-billed Oxpecker (*Buphagus africanus*).

In Kruger National Park in South Africa, C. J. Stutterheim found about one Red-billed Oxpecker (*B. erythrorhynchus*) for every fourteen associated mammals. The sexes were present in equal numbers, and they lived in groups of two to five birds. All members of a breeding unit helped to select the nest site in a tree cavity and to build the nest, almost wholly of hair that they pulled directly from the ungulates, with a small admixture of dry grasses and rootlets. During the incubation period of twelve to thirteen days, the male and female of the single breeding pair covered the two or three eggs alternately by day and the female sat through the night; helpers did not incubate. After the young hatched, the parents at most nests were associated in feeding them by one to three auxiliaries of both sexes, who brought a substantial portion of their meals, removed fecal sacs, and participated in nest defense. Nearly half of the nestlings' food consisted of ticks, supplemented by flies and large amounts of hair, mammalian scurf cells, and epidermal tissue. When the month-old fledglings left the nest, they were led directly to the large mammals, where they continued to be fed by their attendants until they were about three months old and could gather ticks for themselves. Red-billed Oxpeckers can raise up to three broods in a breeding season of nearly six months.

Common Starlings have never, to my knowledge, been recorded breeding cooperatively, but occasionally they serve as interspecific helpers. A pair of starlings feeding their own nestlings in a tree cavity often entered the hole of Northern Flickers (*Colaptes auratus*) in a neighboring tree, apparently fed the flicker nestlings, and carried away their droppings. When a starling fed nestling American Robins (*Turdus migratorius*), their mother continued for a day or two to feed them along with the helper, but after that she was not seen to attend them, probably because she was chased from the nest by the starling. The foster parent continued to nourish the young robins until they left the nest two days later. In Texas, a starling nested in one of the compartments of a house built for Purple Martins (*Progne subis*). After the starling's eggs were destroyed, it began to feed a brood of four young martins in another compartment, bringing them over twenty meals in one hour and also removing their droppings. Disturbed by this unwanted intrusion, the parent martins attacked the helper, who in turn repeatedly drove them from their nest. The martins fed their nestlings in the absence of the starling.

REFERENCES: C. R. Brown 1977; A. Craig 1983; L. G. Grimes 1976; Herbert 1971; Huels 1981; Miskell 1977; Prescott 1971; Stutterheim 1982; Wilkinson 1982.

44.

HONEYEATERS

Meliphagidae

From their center of abundance in Australia and New Guinea, where they are prominent members of the avian community, the 167 species of honeyeaters range widely through the islands of the Pacific Ocean to New Zealand, Samoa, and Hawaii. These very small to fairly large arboreal passerines, 4 to 16 inches (10 to 41 centimeters) long, are variously colored, some displaying brilliant red and yellow, whereas others are largely greenish, brown, or gray. A number of species are adorned with curious projecting tufts of white or yellow feathers. Others have naked cheek patches, wattles, or even featherless heads. Their short to fairly long bills are slender, downcurved, and pointed.

Specialized nectar drinkers, honeyeaters have protrusile tongues with brushy tips and grooved edges. They are among the principal pollinators of the eucalyptus trees that dominate Australian woodlands, and like other nectarivorous birds they tend to be quarrelsome. Honeyeaters vary their diet with pollen and insects; some eat much fruit. Many of the smaller species are excellent songsters, while larger ones draw attention to themselves by vocal mimicry or loud, harsh or bell-like notes freely uttered. In nests that are usually open cups placed in trees or shrubs, rarely domed with a side entrance, they lay from one to four spotted eggs, the number increasing with lati-

Noisy Miner

tude. Although the males usually do not participate in incubation, they nearly always help feed the young.

Although nests of Noisy Miners (*Manorina melanocephala*) frequently have many attendants, I include these birds among the cooperative breeders with certain misgivings, for their boisterous aggressiveness contrasts strongly with the pacific cooperation habitual among birds in this category. Noisy Miners are starling-sized, pale gray birds with yellow-margined wing and tail feathers, white foreheads, black crowns, and whitish underparts. Their short bills, a bare patch behind each eye, and their legs and toes are yellow or yellowish. Inhabitants of open woodlands, they range from Queensland southward along the eastern side of the continent to south Australia and across Bass Strait to Tasmania. Unlike other members of the family, they subsist principally upon a variety of insects gleaned from foliage, bark, and litter. They visited the feeder in front of Ian Rowley's kitchen window at Geary's Gap near Canberra, where he trapped them for banding with bread crumbs fried in fat. Far from being terrified, they helped themselves from the sugar bowl while he fitted bands on them at the breakfast table on cold winter mornings!

The social organization of Noisy Miners differs from that of other well-studied cooperative breeders, which nearly always live in friendly groups that defend their territories from neighboring groups. Nor are they typical colonial nesters. Their colonies, which may contain several hundred individuals, consist of coteries or bands of males who collectively but weakly defend a particular area against other miners and not infrequently enter the space of neighboring bands. Members of these coteries unite to attack with vigor birds of other species, driving them away or occasionally killing them, thereby becoming almost the sole occupants of their living space, an exaggerated form of territorialism rare if not unique among birds. Females, outnumbered by males by as much as 2.2 to 3.5 to 1, are less gregarious, and each tends to remain in her own area. Probably they are so much less numerous than males because they more often emigrate from the colony and confront the perils of unfamiliar terrain, while males tend to remain at home. Among fledglings recently out of the nest, the sexes are represented in equal numbers.

In the population studied by Douglas Dow in southeastern Queensland, Noisy Miners were promiscuous and far more addicted than most birds to conspicuous sexual activity. They roosted in trees where, after much squabbling, six or fewer might perch in contact, as in other cooperative breeders. At dawn, after singing together for about ten minutes, they entered a period of noisy, almost frenzied activity which might last from forty minutes to two hours. Dow wrote: "Violent attacks and chases occurred on all sides. Sometimes feathers were torn out in flight chases. Fights ensued and scuffling birds dropped, claws locked, to the ground. I have seen as many as seven duos of fighting miners drop around me in ten minutes." As one would expect of birds who behave in this fashion, they rarely preen each other.

Only females built the open cups of interlaced twigs and straws, usually high in trees. Males occasionally carried materials, but their contributions to nest construction were negligible. Building females, and males carrying twigs, often approached a nest with a stereotyped head-up display, as though trying to make their activity or the situation of the nest more conspicuous. After laying from two to four eggs, the female incubated for about fifteen days, during which males continued to visit her nest, not, however, to feed her. On the contrary, if the female did not leave when a male approached, he might peck her or pull her from the nest by the feathers of her neck. When the eggs were exposed, he alighted on the rim and displayed to them for six to ten seconds, with yammering, flutter-gaping, and wing waving, all much as in a social greeting display. Lowering his head into the cup, he sometimes touched the eggs with his tongue. If several males arrived at a nest simultaneously, they might fight. Instead of helping make incubation more continuous, as when the female is fed on the nest, the males' visits appeared to disrupt it.

After the eggs hatched, attendants of both sexes continued to approach the nest with the conspicuous head-up display. As the nestlings grew older, the number of attending males, as well as the rate of food bringing, increased until from nine to fourteen males, in addition to the mother, fed a brood of two or three young, sometimes at the surprisingly rapid rate of sixty-three visits per hour. All the attendant males together contributed from 74 to 88 percent of the meals brought to older nestlings. Probably the young were fed so frequently because a meal often consisted of very small insects, scarcely visible in the feeder's bill. The male attendants did not, as in many birds, give food to a brooding female to pass to the nestlings; but each delivered directly what he brought, except when a large item was snatched from the bill of a male in the confusion that arose when a number were present together. Nestlings received their meals with little squeals. Males as well as females removed fecal sacs.

After the young miners left the nest at the age of about fifteen days, both the number of attendants and their rate of feeding increased. Now birds from neighboring groups within the colony appeared to be attracted to the fledglings and fed them if not driven away by the resident miners. A brood of three fledglings received food eighty-eight times an hour—a rate of delivery that few birds approach. Newly emerged fledglings scramble into the surrounding canopy of foliage but may return to their nest to sleep. Within a few days of their departure, the young miners begin to find food for themselves, but they continue to beg from adults and to be fed by them until they are six or seven weeks old. If their family group nests again rather promptly, juveniles may help feed their younger siblings. In the absence of dependent young, these independent juveniles may wander to a neighboring group and feed nestlings there. Yearling miners are among the most active attendants of nestlings and fledglings, contributing from 22 to 29 percent of all feedings at busy nests.

In contrast to the bustling group activity that Dow watched in subtropical Queensland, in an agricultural district farther south, at Geary's Gap in New South Wales, where winter frosts were severe, Ian Rowley watched Noisy Miners for five or six years without finding any evidence that a nest was attended by more than the two parents. The difference between the two populations may have been caused by their varied concentrations. Rowley estimated their density at Geary's Gap at about 10 birds per 50 acres (25 per 50 hectares). In subtropical Queensland they were present at the exceptionally high density of 160 to 200 per 50 acres (400 to 500 per 50 hectares). The difference between the two groups of Noisy Miners provides a striking example of the strong influence of habitat upon patterns of reproduction.

Bell Miners (*Manorina melanophrys*) are found near the Australian coast from southern Queensland to Victoria, where their delightful choruses of bell-like notes ring through forests more humid and dense than those inhabited by Noisy Miners. Like the latter, Bell Miners are mainly insectivorous and live in groups of two to twelve individuals, whose loosely defended areas are aggregated into a colony, seldom far from a stream in wooded country. They may nest in any season, frequently more than once a year. Each group has a single breeding female, who builds her nest and incubates her eggs alone. After the eggs hatch, members of a neighboring group which happens at the time to be without young may help the resident group feed the nestlings, contributing a substantial share of all meals. Since Bell Miners often bring small items, including scale insects gleaned from the foliage of the forest canopy, they feed the young at a rapid rate—up to forty meals per hour for a single nestling, once ninety-two times in an hour for two broods, each of two fledglings, merged into a miniature crèche. In a colony studied in Victoria, Geoffrey Swainson found that groups in the center of the territory nested more successfully than those at its margins, possibly because of more effective defense by members of the colony.

In a more recent study, Michael Clarke demonstrated that although a nonbreeding helper might assist at nests of several pairs, it brought most food to the nest of parents to whom it was most closely related. This correlation between the degree of genetic relationship and the amount of aid that a parent received from nonbreeders provides support for the role of kin selection in the evolution of cooperative breeding in Bell Miners. Male parents feeding their own young not infrequently gave food to offspring of a different pair. At three of five nests, the helpers included not only nonbreeding adults but also juveniles from an earlier brood of the same parents, who alternated between begging or stealing food from other attendants and themselves feeding their younger siblings. Like Noisy Miners, Bell Miners exclude most other kinds of birds from their territory. Clarke concluded that the breeding system of Bell Miners is an adaptation to a generally mesic environment that is occasionally subjected to unpredictable climatic extremes. Multiple small clutches are produced in rapid succession whenever conditions are favorable.

Honeyeaters of the genus *Melithreptus* are small birds that forage and nest in the forest canopy, where they are difficult to band and to study. Our most complete observations on a member of this genus are those made by Richard Noske around his home near Armidale, New South Wales, where Brown-headed Honeyeaters (*M. brevirostris*) were to be found through much of the year in groups of three to eight individuals. In a banded group consisting of a mated pair, two other adults of undetermined sex, and one juvenile, the nest was built by the female alone, accompanied and occasionally fed by her mate. Nest materials included hair that, on repeated visits to a black-and-white calf, the female plucked from its back or side, always from the white patches. After an unstated number of eggs were laid, all four adult members of the group took turns at incubation, the mated male sitting slightly more than his consort, one of the adult helpers somewhat less than either parent, and the other seldom. While sitting in the nest, only the mated female received food from the other three adults, who were never seen to be fed, although some of their sessions of incubation exceeded hers in length. Sometimes, when one of these three came to the nest with food, the mated female flew away before it could be offered to her, leaving the newcomer to incubate and, apparently, eat what it had brought. At another nest, four adults fed two or three nestlings.

To sleep, Brown-headed Honeyeaters, sometimes as many as eight, huddle together in a row on a thin twig or petiole, usually amid foliage near the top of a eucalyptus tree. Roosting individuals face in different directions, possibly to pack more tightly together, as well as to keep watch for the approach of predators from whatever direction. When juveniles are present, they roost in the middle of the row, with two or three adults on each side for their better protection.

Others have watched two nests of Brown-headed Honeyeaters, at each of which three individuals fed a young Pallid Cuckoo (*Cuculus pallidus*). Five adult Golden-backed Honeyeaters (*Melithreptus laetior*) built a nest and later queued up to feed the nestlings. Four adult White-naped Honeyeaters (*M. lunatus*), thought to be two pairs, fed the young at one nest. Years later, three of these honeyeaters fed a fledgling. Another nest held eggs laid by two females, but this species also breeds in simple pairs.

References: Clarke 1984a, 1984b; Dow 1978, 1979; Harrison 1969; Mathews 1924; Noske 1983; Rowley 1975, 1976; Swainson 1970.

45. WHITE-EYES

Zosteropidae

From Africa south of the Sahara across the warmer parts of Asia to Korea and Japan, in Australia, New Zealand, and on islands widely scattered through the Indian and Pacific oceans, ninety species of white-eyes flourish. These small songbirds, 4 to 5.5 inches (10 to 14 centimeters) long, differ surprisingly little in appearance. Their upperparts are olive-green, grayish, or brownish; their underparts white, gray, or yellow. Males and females are dressed alike, and all lack bright plumage. White-eyes receive their name from the prominent ring of minute, silky, white feathers around each eye; only in one species is the eye ring yellow. Flocking through woodland, thickets, arid scrub, or mangrove swamps, they search the foliage for small insects, including many aphids and scale insects, gather berries and other fruits, or with short, sharp bills and brush-tipped protrusile tongues extract nectar from flowers—sometimes by piercing the bases of tubular corollas, as hummingbirds frequently do. Their prolonged, richly varied songs are often loud and far-carrying.

The breeding pair builds a deep cup of vegetable materials attached, vireo-like, between the arms of a forked twig or between a stout petiole and stem. Both sexes incubate the two or three, rarely more, white or blue, immaculate eggs, which hatch in ten and a half to twelve days, and both nourish the nestlings with tiny caterpillars and small insects. The young of three African species left their nests when thirteen to fifteen days old. Some white-eyes permit themselves to be touched while they sit on their nests and take food from a human hand.

White-eyes are highly social birds, foraging and bathing in flocks, often building their nests close together, and sleeping in communal roosts in leafy trees, where the members of a pair perch in contact but separated from other pairs. After the young fly, the family snuggles together on a twig at night, the fledglings in the center with a parent on either side. Mates feed and preen each other. One expects birds so social to nest cooperatively and, strangely enough, confirmation comes from remote islands in the Indian Ocean rather than from continents where white-eyes abound. On the Seychelles Islands, P. W. Greig-Smith watched four Seychelles White-eyes (*Zosterops modesta*), all alike in appearance, build a nest together, each bringing materials and usually arranging its own contribution. At another nest, four individuals shared incubation, sitting from seven to fifty-two minutes at a stretch. Observations on the care of young are lacking. The relationship of the helpers was not determined, but they may have been the breeding pair's offspring, hatched earlier in the same year. Two, three, or four birds of this group were repeatedly seen huddled in close contact, preening each other's heads and necks or pecking at their white eye rings. On the Mascarene Islands, Frank Gill found Mascarene White-eyes (*Z. borbonica*) nesting in loose colonies, where helpers sometimes fed the young.

Seychelles White-eye

White-eyes are well known as successful colonizers of remote islands, whether they reach them by their own efforts or are brought to them. Since the introduction of the Japanese White-eye (*Z. japonica*) to Hawaii in 1929, it has spread widely over the archipelago. When Robert Eddinger placed fledgling Japanese White-eyes in a large cage with a slightly older juvenile that he had hand-fed until it could feed itself, it immediately brought food to the begging young. When Eddinger entered the enclosure, the helper alighted on his finger, accepted food from him, and carried it to the fledglings. Later, when he had four immature birds, they contended for the privilege of feeding a newly acquired fledgling. These young white-eyes also fed nestling House Finches (*Carpodacus mexicanus*) and House Sparrows (*Passer domesticus*) deposited in their cage, continuing until their fosterlings no longer needed their ministrations. It was remarkable how promptly the juvenile white-eyes responded to the begging of any helpless birdling newly set among them. With such eager fosterers, Eddinger no longer needed to hand-feed the callow young in his aviary. These young white-eyes frequently preened one another. They slept, and often perched by day, pressed so closely together that they looked like one bird with five heads.

References: Eddinger 1967, 1970; Gill 1971; Greig-Smith 1979; Skead and Ranger 1958; Van Someren 1956.

46. WOOD-WARBLERS

Parulidae

The 117 species of wood-warblers are confined to North America, South America, and the Antilles. Long considered a separate family, they are now sometimes classified as a subfamily (Parulinae) of the huge, unwieldy emberizine assemblage. Here we follow the earlier treatment. Slender-billed, dainty birds from 4 to 7 inches (10 to 18 centimeters) long, they are red, orange, yellow, blue, chestnut, black, white, and olive in the most varied patterns. Most flit through trees and shrubbery seeking insects and spiders; some dwell amid low, dense herbage; a few walk or hop over the ground. Fruit is a minor constituent of the diets of some. Species that nest where winters are severe undertake long nocturnal journeys to warmer lands. The more ornate males of these migratory wood-warblers often have plainly clad mates, and they molt into duller plumage after the breeding season. In the many species permanently resident in tropical America, chiefly at middle and high altitudes, the sexes tend to be alike and to wear the same colorful plumage throughout the year. Wood-warblers build open or roofed nests in trees or shrubs or on the ground; they lay two to four or five, rarely six, white or tinted, spotted eggs that are incubated by the female alone. Both sexes feed the young.

Since the males are zealous, songful, monogamous defenders of territories, and many are migratory, wood-warblers are not likely to breed cooperatively. As in other strongly territorial birds, instances of helpfulness are more often found in interspecific than in intraspecific contexts. Most recorded instances of intraspecific helpers were disclosed by Val Nolan's prolonged studies of the Prairie Warbler (*Dendroica discolor*) in Indiana. A female who had lost her mate was joined by a male from an adjacent territory who had lost his mate and nestlings. He fed her nestlings

Prairie Warbler

and cleaned the nest for three days. Then he lost interest, as it was late in the season, when some males were neglecting even their own offspring. Another male, likewise bereft of mate and offspring, also fed neighboring nestlings. Eight males carried food into thickets where young Prairie Warblers not their own were begging. Five of the eight were trespassing on the territories of the fledglings' fathers; two had incubating mates. The other three altruistic males were on their own territories, into which fledglings not their own were led by their mothers. Some males promptly fed alien young introduced into nests where their mates were incubating eggs, and adults of both sexes fed fledglings experimentally placed in their territories.

At a nest of the Kentucky Warbler (*Oporornis formosus*), a second male, who seemed to be abnormal, brought food to the incubating female, in spite of being chased by another male who sang better and appeared to be her mate.

In New York State, a banded male Blue-winged Warbler (*Vermivora pinus*) brought food to nestlings of an established pair. A week after these young fledged, he was seen carrying food near a nest with young of a different pair. In the following year, another male helper was found attending nestlings of an established pair. Like the first of these helpers, he later fed a different brood. In another locality in New York, three Blue-winged Warblers fed two fledglings. Helpers, apparently mostly unmated males, seem to be not uncommon among Blue-wings. Sometimes the situation becomes complicated. In Connecticut, a pair consisting of a Blue-winged Warbler mated to a hybrid between this species and the Golden-winged Warbler (*V. chrysoptera*), known as Brewster's Warbler, was assisted in feeding its nestlings by another hybrid between these two species, the form called Lawrence's Warbler. The first male chased and fought with the second.

Among interspecific helpers in this family was a Black-and-white Warbler (*Mniotilta varia*) who repeatedly fed nestling Worm-eating Warblers (*Helmintheros vermivorus*). The parents attacked him when he approached their nest, and once they tore food from his bill and themselves gave it to their brood. The helper gave a distraction display when the observers visited the nest. For several days, a pair of Black-and-white Warblers fed a recently fledged Ovenbird (*Seiurus auricapillus*) who had apparently become separated from its parents. A nestful of Ovenbirds was for at least five days fed by a Worm-eating Warbler. When the helper and the nestlings' mother were at the nest together, they evinced no hostility. An American Redstart (*Setophaga ruticilla*) assisted at a nest of the Yellow Warbler (*Dendroica petechia*). While young redstarts were being photographed in the hands of children, their father brought food to them, but their more timid mother presented the contents of her bill to a brood of American Robins (*Turdus migratorius*) in a nest 25 feet (8 meters) away. Very exceptionally, two wood-warblers share a nest, as happened when two female redstarts laid together; both attended the double set of eggs, while two males remained nearby.

REFERENCES: Allen 1930; Bent 1953; De Garis 1936; Kendeigh 1945; Maciula 1960; Nolan 1978; Rea 1945; Short 1964.

47.

TANAGERS AND HONEYCREEPERS

Thraupidae

The 230 species of tanagers and honeycreepers spread widely over the continents and islands of tropical America, from lowlands to high in the mountains, but only four migratory species are found north of Mexico. Clad in all the brightest colors as well as olive, gray, brown, and black, they contribute more than any other family to the splendor of tropical American birds. With song, however, they are on the whole poorly endowed; a few sing sweetly or persistently, but many of the most brilliant are nearly or quite songless. Largely frugivorous, they vary their diet with insects and nectar. Their nests, neat open cups or, among the euphonias and chlorophonias, globular structures with a side entrance, are built by both sexes or, in many species, by the female alone. As far as is known, only the female incubates the two to five eggs, sometimes while she is fed by her mate, who nearly always helps her attend the nestlings.

Among so many brightly colored species, the Dusky-faced Tanager (*Mitrospingus cassinii*) is an anomaly. A slender bird nearly 7 inches (18 centimeters) long, it has a grayish black forehead, face, and chin. Elsewhere it is dark grayish olive and yellowish olive-green. Its rather long bill and dull yellow eyes are rare among tanagers. From the wet Caribbean lowlands of

Costa Rica it ranges through Panama and the rainy Pacific lowlands of Colombia to northwestern Ecuador. With a constant chatter of sharp, harsh notes and apparently never a song, family groups of six or eight, rarely ten, individuals seek berries and insects while they wander restlessly through low, dense vegetation at the margins of rain forests, along woodland streams, about the edges of swampy openings, and in nearby thickets and shady plantations. When I first met these birds, they reminded me more of jays than of tanagers, and I at once suspected that, like many jays, they are cooperative breeders.

To find a nest of these tanagers that frequent thickets where a man can hardly advance without cutting his way with a machete was not easy; in two seasons in northeastern Costa Rica we discovered only one. Situated about 10 feet (3 meters) up in a small tree beside a forest stream, the bulky open cup, composed almost wholly of long, brown, threadlike, dry pistillate inflorescences of a small tree (*Myriocarpa*), was slung between two upright branches a few inches apart, with no support below. Knowing how precarious is the life of a nest in the tropical forest, we did not jeopardize it by cutting a path to it but watched from the high bank 50 yards (46 meters) away across the stream, from which we could see two nestlings when they stretched up their gaping mouths with bright red interiors. In six hours of a morning, they received forty-three meals, each a single insect held in an attendant's sharp bill. Of the six or seven Dusky-faces that flitted noisily through the tangles around the nest, at least three fed the nestlings. Probably more brought food, but their reluctance to expose themselves more than momentarily to a watcher—even one partly screened by foliage across the stream—made it difficult to ascertain this. Later, I saw three of these tanagers carry food into a dense thicket along a different stream, for nestlings or fledglings that I could not find.

A more typical tanager is the Golden-masked, or Golden-hooded (*Tangara larvata*), a 5-inch (13-centimeter) bird clad in an intricate pattern of golden-buff, blue, turquoise, black, and white. Throughout the year it lives in pairs or groups of three or four in the upper levels of rain forests, clearings with scattered trees, and shady plantations and gardens. A male and female, working together, build a compact open cup in a tree or shrub, where the female, fed occasionally by her mate, incubates two white or pale gray eggs thickly sprinkled all over with brown and chocolate.

At one nest in Panama and three in Costa Rica, I watched a Golden-masked Tanager in juvenal or transitional plumage help feed nestlings of a later brood, probably always their younger siblings. One of these young helpers engaged in this activity when forty-six days old. Often the juvenile flew to the nest with the two parents, but sometimes it came alone. It brought smaller items than the parents did and its attendance was erratic, with intervals of coming about as frequently as the adults alternating with periods of neglect. Occasionally, with quivering wings, the helper solicited food from the parents, and it carried away fecal sacs. On several occasions, while their parents built a nest for a sec-

Golden-masked Tanager

ond brood, I have seen juvenile Golden-masked Tanagers make gestures of helpfulness—such as gathering materials, which they soon dropped, or making shaping movements in the bowl—but their contributions to the undertaking were negligible.

In three localities, I have watched trios of adult Golden-masked Tanagers, in full breeding plumage, attending nestlings. One of these trios brought food to a nest that held one nestling and one unhatched egg, so evidently it was not a case of two females laying in the same nest. One September, from my study window, I watched four adults repeatedly carry food to a late nest. They tended to come in pairs rather than as a flock, but when all were present together I noticed no antagonism between them. Unfortunately, I could not learn how many young this inaccessible nest contained. Nor could I ascertain the relationship or the sexes of the three or four adult attendants at a nest, since males and females wear the same elegant attire.

A pair of Speckled Tanagers (*T. guttata*) built in a poró tree (*Erythrina berteroana*) near our house. The female was lame, with a right leg that hung uselessly, yet despite this she hatched a single nestling, who was fed by two sound Speckled Tanagers in addition to its mother. I could not distinguish the helper from the lame female's mate. In nine hours, she brought food fifty-two times and the other two attendants sixty-two times. Although sometimes all three came to the nest together, more often the helper came alone. After the fledgling left the nest, I repeatedly saw its three attendants carry pieces of banana from the feeder. Between breeding seasons, Speckled Tanagers often fly in groups of four or five.

Speckled Tanager

The Plain-colored Tanager (*Tangara inornata*), an unadorned gray member of an ornate genus, inhabits woodlands from Costa Rica to Colombia. In a narrow clearing amid the forests of Panama, I watched four grown birds, all alike in appearance, feed two nestlings, often arriving in a little flock.

The blue, black, turquoise, and yellow Turquoise Tanager (*T. mexicana*) is widely spread over South America from Colombia and Venezuela to Bolivia and

Turquoise Tanager

Brazil. On the island of Trinidad it lives in groups of four to seven, rarely more, and hardly ever forages alone. On a number of occasions, four or five adults were seen attending nestlings or feeding fledglings. Apparently, mates separate themselves temporarily from their associates to build and incubate, but after the eggs hatch the whole group cares for the young until they attain independence.

Red-throated Ant-Tanagers (*Habia fuscicauda*) live in family groups at the margins of forests and in second-growth woods, plantations, and mangrove swamps from southern Mexico through much of Central America to northern

Colombia. They lay three white eggs in open cups that the female builds in shrubs or small trees of the woodland undergrowth. Occasionally a nonbreeding bird brings materials and works on the nest. Adult females other than the mother and also juveniles bring food to the nestlings, one of the juveniles about as frequently as either parent did; but such assistance appears to be rare. A male repeatedly fed his neighbor's two fledglings, in spite of being persistently driven away by their father.

In Costa Rica, a full-grown young Scarlet-rumped Tanager (*Ramphocelus passerinii*), afflicted with paralysis, was fed at least once by each of two different females, one of whom appeared to be its own parent, while the other was the mother of fledglings just out of the nest.

At their home close by a Panamanian forest, Roger Johnson and Lorraine Washington watched Thick-billed Euphonias (*Euphonia laniirostris*) build globular nests with a side entrance in niches that the birds dug into the soil of hanging baskets of ferns. A nest with five hatchlings was frequently visited by three or four euphonias who appeared to feed them. Traces of immaturity in the plumage of some of the male attendants suggested that they were yearling sons of the pair whose nest they visited. In the following year, with three occupied nests in neighboring baskets and at least ten grown birds showing interest in them, the situation became too confusing to learn how many visited any one nest. I have watched one or more nests of five species of euphonias without finding any indication of helpers, but the Thick-billed Euphonia appears to be different and would repay prolonged study.

Among the more widespread and familiar of the tanagers is the Blue-gray (*Thraupis episcopus*), which from central Mexico to Peru, Bolivia, and Amazonian Brazil frequents city parks and suburban gardens as well as farms, open woodland, and the forest's edge. Throughout the year it lives in almost inseparable monogamous pairs; but once, near our house in Costa Rica, I found a bigamous male and two females attending a nest, which contained four eggs instead of the two that are almost invariably laid in southern Central America. The dominant female, in better plumage than the submissive female, incubated when she wished; the latter sat in the nest the moment the former left. If the submissive female was slow in vacating the nest when the other returned from a recess, she was dismissed with a mild peck. The two never incubated simultaneously, but together they kept the eggs constantly covered, which is unusual in the tanager family. After the nestlings hatched, the male and two fe-

Thick-billed Euphonias, adult male (left), female (center), and young male in transitional plumage

males fed them in perfect harmony, bringing 142 meals to the three nestlings in four hours. At least two young fledged.

I have found only two records of tanagers feeding birds of other species. Both of these interspecific helpers were males of the highly migratory Scarlet Tanager (*Piranga olivacea*), which breeds east of the Rockies in Canada and the United States. One fed nestling Chipping Sparrows (*Spizella passerina*) until his own young hatched. Many years later, another brought food to nestling Chipping Sparrows, whose parents tried to drive the helper away.

From the foregoing survey, it appears that helpers occur sporadically in a number of weakly territorial tanagers but cooperative breeding is rare in the family. In only two or three of the species that have been studied—the Dusky-faced and Turquoise tanagers and possibly the Speckled Tanager—is it likely to be found at a substantial proportion of nests, but we need many more observations to establish this. The great majority of tanagers await patient study by those who can find their well-hidden nests.

In the American Ornithologists' Union's latest *Check-list of North American Birds*, the heterogeneous honeycreeper family (Coerebidae) has been dismembered, and four of its genera (*Conirostrum*, *Chlorophanes*, *Dacnis*, and *Cyanerpes*) have been transferred to the tanagers, there reduced to the status of a subfamily (Thraupinae) of the Emberizidae. These lovely little birds appear to be tanagers that have evolved thinner, more pointed, and often conspicuously longer bills for probing flowers for nectar, which thick-billed tanagers do only occasionally and less expertly. Honeycreepers also eat many berries and arillate seeds and are attracted to feeders where fruits are displayed. Their rare attempts to sing are disappointing. They build slight, open nests in trees and shrubs and lay two eggs that are incubated by the female alone.

At one of the two nests of the Blue Dacnis (*Dacnis cayana*) that I have seen, the male sometimes fed his mate while she incubated her speckled eggs. A second male, also in full nuptial attire, was occasionally nearby. After one of the eggs hatched, both males helped the female feed the nestling. Sometimes one male mildly chased the other, yet at other times they were at the nest together with food. When I inspected the high nest with a mirror raised upon a long pole, both protested my intrusion along with the female, all voicing low, weak notes. Although this may have been a case of polyandry, I think it more probable that the second male was a helper rather than another mate of the female. One of the males associated with her much more closely than the other.

One day in April, my wife saw a male Red-legged, or Blue, Honeycreeper (*Cyanerpes cyaneus*) in full breeding plumage pass food to a fledgling Scarlet-rumped Tanager on our feeder. For the next three days, the brilliant honeycreeper continued to attend the tanager, who was about twice his size, giving it chiefly pieces of banana or plantain from the feeder and sometimes insects caught among the foliage. Again and again, this strange pair returned to the feeder, where the honeycreeper stuffed the tanager with fruit, once passing it six billfuls in rapid succession. The honeycreeper insisted upon pushing his long, sharp bill well down into the throat of the short-billed tanager, who seemed not to relish this method of delivering food. When sated, the young tanager would turn its head away, whereupon the honeycreeper would flit over its back from side to side, presenting morsels alternately on the right and on the left, until the tanager flew away with its attendant following. Often the young bird pursued the honeycreeper through the neighboring trees, begging; but when the attendant started on a high flight, the tanager, whose kind does not fly so high and far, did not follow. The fledgling was beginning to feed itself, and it also received at least occasional meals from an adult male Scarlet-rumped Tanager who appeared to be its father. It seemed, however, to prefer the gifts of the more complaisant honeycreeper. No female tanager was seen to feed the fledgling.

For at least two days in a later year, a female Red-legged Honeycreeper gave billfuls of banana from our feeder to a fledgling Yellow-green Vireo (*Vireo flavoviridis*).

References: Hales 1896; R. B. Johnson *in litt.*; Prescott 1965; Skutch 1954, 1961, 1962, 1972, 1980, 1981; Snow and Collins 1962; Willis 1961.

48.

NEW WORLD BLACKBIRDS AND ORIOLES

Icteridae

The eighty-eight species of this family are an exceedingly heterogeneous assemblage, confined to the woodlands, thickets, meadows, and marshes of the western hemisphere, from the Arctic tundra to Tierra del Fuego and from lowlands to high in the mountains. Collectively known as icterids, they include orioles, oropendolas, caciques, grackles, blackbirds, marshbirds, meadowlarks, the Troupial, and the Bobolink. In size they range from 6 to 21 inches (15 to 53 centimeters) and in plumage from plain black to brilliant yellow or orange, set off by black and white. Even the largely frugivorous species take many insects, and some are almost omnivorous. Their notes, no less diverse than their appearance and habits, include bright melodious songs, liquid gurgles, dry rattles, and shrill screams.

The most skillful avian weavers in the New World, the orioles and oropendolas hang their long pouches high in trees, singly or in crowded colonies. At the other extreme, Bobolinks (*Dolichonyx oryzivorus*) hide slight open cups at the roots of northern meadow grasses. It is strange that a family which contains so many accomplished builders should include the parasitic cowbirds, which make no nests but drop their eggs into those of other birds that serve as foster parents. In other icterids, the female alone incubates from two to six, rarely more, eggs, which range in color from

Bay-winged Cowbird

white to greenish or blue, with or without markings. Males of the monogamous species usually help feed the nestlings, but in the polygamous, colonial species they often do little more than sing, display, and guard the nests.

Surprisingly, one of the few members of this family known to breed cooperatively is the only nonparasitic cowbird. The Bay-winged Cowbird (*Molothrus badius*), an 8-inch (20-centimeter) grayish brown bird with rufous-chestnut wings, inhabits South America from tropical Brazil far southward into the temperate zone of Argentina. Social at all seasons, these cowbirds live in flocks of ten to thirty individuals, who often sing together, pleasantly but not brilliantly, in a subdued, leisurely manner. For nesting they prefer the bulky structure of interlaced sticks built by Firewood-Gatherers (*Anumbius annumbi*) or Brown Cachalotes (*Pseudoseisura lophotes*) but will occupy other closed structures, such as the domed nests of the Great Kiskadee (*Pitangus sulphuratus*), the clay ovens of the Rufous Hornero (*Furnarius rufus*), nest boxes, or woodpecker holes. Usually they choose nests no longer occupied by the builders, but if necessary they will fight stubbornly for possession of one. Occasionally, especially early in the season when few nests of other birds are available, they build a crude open cup in a crotch or cranny in a tree.

The female Bay-wing selects the nest site, lines the borrowed structure, sleeps in it before she lays her four or five whitish eggs, heavily marked with reddish brown, incubates them for twelve days, broods and feeds the nestlings, and feeds the fledglings. Her mate at first follows her passively but later may help build; he does not incubate but feeds the young before and after they leave the nest, at the age of fourteen or fifteen days. He also guards the nest and chases the Screaming Cowbirds (*Molothrus rufoaxillaris*) that persistently try to parasitize it, for they have no other host.

A population of Bay-winged Cowbirds studied by Rosendo Fraga in the Buenos Aires Province of Argentina consisted of fourteen males and eight females. At five of the eight nests in which eggs hatched, one or two extra males helped the parents feed and guard the nestlings. In all cases, as soon as the young left the nests, the number of their attendants increased when wandering cowbirds joined the parents and original helpers in feeding the fledglings and mobbing predators. One young family was surrounded by a noisy, excited crowd of no fewer than eighteen adults—too many to ascertain that all were bringing food. When the Bay-wings have raised Screaming Cowbirds, the fosterlings, which closely resemble the legitimate offspring, receive similar attention. After raising his own brood, a male Bay-wing may help at the nest of another pair. In Fraga's study area, no breeding pair failed to receive aid, after their young fledged if not before. In the province of Catamarca in northwestern Argentina, Gordon Orians and his family found more than two adults at all three of the nests with young that they watched.

Brown-and-Yellow Marshbirds (*Pseudoleistes virescens*), which W. H. Hudson called Yellow-breasted Marshbirds, reside from extreme southern Brazil through Uruguay to central Argentina. About 9 inches (23 centimeters) long, both sexes are glossy olive-brown with bright yellow shoulders, lower breast, and abdomen. Sociable and noisy, these marshbirds live throughout the year in flocks of from twenty to thirty individuals, who forage on the ground in fields and wet pastures, with one bird serving as a sentinel while its companions eat. After feeding, they settle in trees where they join their voices in an artless chorus that Hudson called a "delightful hubbub." Peaceable birds, they appear rarely to quarrel among themselves or with birds of other kinds. In a low tree or dense shrub or amid tall grasses, cattails, or sedges, the female builds a thrushlike nest of dry grasses and fine twigs, plasters it inside with mud or cow dung, and lines it with hair or soft grasses. Here she lays, and incubates alone, from three to five white eggs spotted with deep red, very heavily on the thicker end. Often these nests are as little as 4 yards (4 meters) apart, as Brown-and-Yellow Marshbirds do not defend territories.

While the females built, each was accompanied only by her mate. While they incubated, one or two marshbirds in addition to their mates fed them at six of the ten nests that Orians and his team watched during this period. After the fledglings hatched, the number of helpers increased, until at eight of twelve nests from one to six of them, most often two or three, joined the two parents in bringing food to the young. Neither the sexes of the auxiliaries nor their relationship to the parents could be learned during a study that continued for somewhat less than two months, but despite the high predation on nests, the helpers did not appear to be birds who had lost their own broods.

The male of the 11-inch (28-centimeter) Austral Blackbird (*Curaeus curaeus*) is everywhere glossy black, the female browner and duller. In southern Argentina and Chile, these large blackbirds flock in pastures and openings amid forests of southern beeches (*Nothofagus*). Orians and his group found from four to six birds so closely associated that they appeared to breed cooperatively. Unfortunately, the party of observers arrived too late in the season to find nests, but they saw a fledgling receive food from at least three of the four adults who attended it closely, and in another locality they watched six adults escorting a young blackbird.

In the mountains of Bolivia, 9-inch (23-centimeter) Bolivian Blackbirds (*Oreospar bolivianus*) live in flocks of about four to eight in canyons walled by steep cliffs. They forage on the ground amid the bunchgrasses whose seeds they eat, and they build cup-shaped nests in deep crevices in the cliffs. When a female left the nest where she was incubating three greenish gray eggs spotted with gray and brown, she flew to her group, begged from her companions, and was promptly fed by them. Orians and his friends saw a single blackbird receive food five times in one minute, from four different flock members. They never saw a bird with food refuse to deliver it to a begging companion. When she returned to her nest, the incubating female was sometimes accompanied by two other blackbirds. These observations, made in a six-day study, are hardly adequate to prove that Bolivian Blackbirds are cooperative breeders, but they point strongly to this conclusion.

In a colony of Common Grackles (*Quiscalus quiscula*) in the botanical garden of the University of Michigan, Henry Howe found most birds breeding in monogamous pairs and detected only a few nest helpers. A male who had lost his brood to a predator fed young in a neighboring nest but was not joined in this activity by his mate. At another nest, two males brought food throughout the nestling period. When they met at the nest, neither was antagonistic toward the other, and the nestlings' mother did not discriminate between them. In this colony, as many males were helpers as were polygynous. In Quebec, Canada, a female Common Grackle fed and protected three nestling Chipping Sparrows (*Spizella passerina*) whose parents could not be found—unexpected behavior of a bird that sometimes preys upon other birds and their nests.

In Minnesota, a female Red-winged Blackbird (*Agelaius phoeniceus*) who lost her fledglings two days after they left the nest helped for the next ten days to feed the fledglings of a more fortunate neighbor. Only a few isolated incidents of helping at the nest have been disclosed by the many prolonged studies of North American icterids, while in South America far less fieldwork has brought to light a number of species in this family that breed cooperatively, and almost certainly more remain to be discovered.

And not only in the southern continent! Soon after I had written the foregoing, I read a paper by Robert Beason and Leslie Trout, who found helpers at three of fourteen nests of the Bobolink in the state of New York. At one of these nests, two females and one male carried food to five five-day-old young. At another nest, one female and three males simultaneously attended five nine-day-old nestlings. At the third nest, at least two adult females and one adult male brought food to four ten-day-old young. These discoveries were made incidentally at nests kept under observation for another purpose. One hardly expects adult helpers in a long-distance migrant like the Bobolink; but the high site fidelity of these long-lived icterids suggests the possibility that the helpers were offspring, hatched in an earlier year, of the parents at whose nests they assisted. Or could they have been breeding adults who, after the loss of their own brood, satisfied their parental impulses by attendance at nests of more fortunate neighbors?

References: Beason and Trout 1984; Bent 1958; Fraga 1972; Howe 1979; Hudson 1920; Orians et al. 1977a, 1977b; Strosnider 1960.

49.

GROSBEAKS, FINCHES, AND SPARROWS

Fringillidae

Recent years have seen disconcerting changes in the classification of passerine birds. Old, familiar families have been split or lumped in huge, unwieldy families. Since this is not a taxonomic work, in this chapter I shall disregard recent changes in taxonomy and consider together a number of species of the old finch family, the Fringillidae, although many are now included in the Emberizidae. In any case, the prevailing strong territoriality of individual males of these songful birds is hardly compatible with cooperative breeding or the habitual presence of nest helpers. With two exceptions, the birds of which I shall tell are only casually or more or less accidentally helpers, often in interspecific contexts.

An exceptional species is the Black-faced Grosbeak (*Caryothraustes poliogaster*), a stout, 8-inch (20-centimeter) finch with a black patch surrounding the base of the short, very thick bill and extending to the forepart of the cheeks and the upper throat. The rest of the head, the neck all around, and the breast are yellow. Elsewhere, this not very brightly colored bird is olive-green and gray. The sexes are indistinguishable. From southern Mexico to western Panama, this grosbeak wanders in straggling flocks of a dozen or two individuals through the upper levels of Caribbean rain forests and clearings with scattered tall trees, including shaded cacao plantations and pastures, in the lowlands and foothills up to about 3,000 feet (915 meters) above sea level. It eats a variety of small fruits of trees and shrubs and—with slow, deliberate movements—hunts insects and their larvae amid the foliage of trees. A loquacious bird, it utters a variety of dull or buzzy notes, a scarcely melodious *chip chip chip*, and more rarely a pleasant song of about six notes.

Nests of Black-faced Grosbeaks seem usually to be hidden amid the epiphytes that grow profusely on trees of humid forests. For two seasons, my son and I searched for nests amid such forests in northeastern Costa Rica but found only one. This, however, was most conveniently situated 9 feet (2.7 meters) up in a small, sickly tree, profusely burdened with aroids, bromeliads, gesneriads, orchids, ferns, and mosses, right behind our dwelling in a narrow clearing in the forest. We watched the male and female, chiefly the latter, build a shallow, flimsy bowl with materials that they gathered from trees rather than the ground. They worked in a most desultory fashion, bringing a few pieces, then disappearing for a long while, with the result that twelve days elapsed between the beginning of construction and the start of laying. The set consisted of three dull white eggs, mottled and spotted all over with bright shades of brown, chiefly on the thicker end.

Although we could not with certainty distinguish the sexes, prolonged watching convinced us that only the female incubated. She took long sessions, from half an hour to two hours, but her recesses were also rather long. Her return to the nest was the occasion of a spectacular ceremony. Coming from a distance, the pair perched high in a tree at the edge of the clearing. Then both flew sharply downward toward the little nest tree, close together, as though they raced each other to this goal. The female alighted in the nest tree, while her partner swept past to rise into the tall trees at the other side of the clearing, describing a deep catenary loop between opposite edges of the forest. This was repeated many times and may have misled a predator that failed to notice that one bird remained in the nest tree while the other flew past it more conspicuously. During building and well into the incubation period, we had never noticed more than two grosbeaks at this nest so easy to watch over. But a few days before the eggs hatched, two grosbeaks were regularly escorting the female to her nest, exactly as one had formerly done. Quite similar to the mated pair in appearance, the newcomer's sex and relation to them remained unknown. Occasionally one of these companions fed the other or gave food to the incubating female.

When, after an incubation period of thirteen days, the pink-skinned, sparsely downy nestlings hatched, they were fed by this helper in addition to the parents. Until, at the age of twelve days, the feathered young left the nest and soon passed from our ken into the surrounding rain forest, these three fed them regularly, usually coming and going together after brooding ceased; sometimes two of them fed simultaneously from opposite sides of the nest. We never noticed the least antagonism between them or any indication of dominance. Occasionally during the nestling period, the three attendants were followed by two to four other grosbeaks, who flitted around the nest and sang, mostly without any sign of discord; once, however, when four individuals were present, one chased another mildly, but not out of the sparsely branched tree. Twice we saw four birds feed or appear to feed the nestlings, but only the original three attended them consistently.

A decade after we made these observations, Timothy Moermond found, in a bromeliad perched on the trunk of a spiny palm tree in the same locality, the second recorded nest of the Black-faced Grosbeak. It already held nestlings that four adults were feeding. Sometimes six birds came near the nest, but not all were seen to bring food. When Moermond caught and held a fledgling that had just fluttered from the nest, four adults darted close around him, protesting.

Although observations at only two nests may not seem sufficient to allow us to include Black-faced Grosbeaks among habitual cooperative breeders, all that we know about their behavior points strongly in this direction. Even in the nesting season, at least from April to July, they roam through the trees in flocks. They use their voices freely but sing rather sparingly and apparently not to proclaim possession of territory; they do not chase other grosbeaks from near their nests. From among the flocks of nonbreeders that wander about during the breeding season, individuals with stronger motivation join, as regular helpers, a breeding pair, who may be their parents; others bring only an occasional contribution of food, then fly away with their companions.

On Daphne Major Island in the Galápagos Archipelago, Trevor Price and his colleagues watched eleven nests of the Cactus Finch (*Geospiza scandens*). At each of these nests, the young were fed by at least one helper, who was an unpaired, territory-holding, adult male at least three years old. Four of these nests had two helpers, and each of four helpers visited two nests alternately. Yearling and two-year-old males were not found helping. If the parents encountered their assistants away from the nest, they chased them; if they found the same individuals already at the nest, these same parents usually permitted them to feed the nestlings. These bachelor males apparently served as helpers because females were

Black-faced Grosbeak

scarce and they could not find mates; by the following year, all but one of the eleven helpers had paired. One Cactus Finch served at a nest of the Medium Ground-Finch (*G. fortis*). Moreover, at two of the six nests of the latter that were watched for full days, a helper of the same species was present.

At a nest of the Hoary Redpoll (*Carduelis hornemanni*) in the Canadian Arctic, Fred Alsop and his companion saw three males bring food to a female brooding her nestlings. One of these males was in full adult plumage, and two were not much brighter than the female. At another nest in the same neighborhood, at least two males delivered food to the brooding female. The relationship of the males to the females whom they fed was not known.

Male Smith's Longspurs (*Calcarius pictus*) make little effort to defend their song perches, mates, or nest sites on the Arctic tundra. At one nest, the young were fed by two pairs of adults. The helpers' own nestlings had apparently succumbed during a storm.

Rarely, two females of the same species of sparrow or bunting lay in the same nest and together incubate their eggs and feed their young. Among these aberrant females were two Song Sparrows (*Melospiza melodia*) who incubated their double clutch alternately. After the eggs hatched, all four parents (the two females and their respective mates) fed the eight young. I have found three records of two female Northern Cardinals (*Cardinalis cardinalis*), one of whom was an albino, laying in the same nest, incubating side by side, and cooperating in feeding the nestlings if the doubly incubated eggs hatched. In England, two female Reed Buntings (*Emberiza schoeniclus*) shared a nest; they and one male fed the nestlings. The male of this bunting is occasionally bigamous, with two females who build separate nests in his territory. Sometimes he helps feed both broods, but more often he neglects them.

Altruism was exhibited by a male Dark-eyed Junco (*Junco hyemalis*) who helped at the nest of a different pair. On the morning when a nestling of this pair hatched, its father was found dead. In the afternoon of the same day, the widowed mother was joined by the altruistic male, who four days after his arrival started to feed the nestling and continued to bring food during the five additional days that the young junco remained in the nest. He contributed more than half of the observed feedings. Both adults fed the fledgling until it was at least fourteen days old and had been out of the nest for four days, and both actively defended the nest with alarm calls and distraction displays. Likewise altruistic were two Palilas (*Loxioides bailleui*), a Hawaiian finch, that helped the male parent feed nestlings whose brooding mother had been killed by a cat. A semicaptive male Rose-breasted Grosbeak (*Pheucticus ludovicianus*), passing with food for his own nestlings, responded to the loud calls of the young in a neighboring nest, whose parents neglected them, and fed them repeatedly.

Nests of the large, plain-colored Buff-throated Saltator (*Saltator maximus*), widespread in tropical America, are usually well separated, but once in a Costa Rican plantation I found two in coffee bushes only 8 feet (2.4 meters) apart. Each belonged to a monogamous pair. Defending her territory, the dominant female repeatedly chased her neighbor from the latter's eggs and nestling. At intervals the dominant bird visited the other's nest to make sure that her rival was absent. If she happened to be carrying food for her own nestling, she might give it to the offspring of the female she was trying to drive away. Twice I saw her feed her neighbor's nestling, and once she brooded it for five minutes. This episode illustrates the strength of birds' parental impulses, which can overcome territorial aggressiveness. To kill the subordinate female's progeny would have been the quickest way to effect her departure.

Although grosbeaks, finches, and sparrows often raise two or more broods in a season, young from an early brood appear rarely to help at a later nest. Among the few examples of a juvenile helper among these birds that I have found was a Northern Cardinal, probably a female about seventy-eight days old, who brought food to a late nest of her own species. She begged from the nestlings' parents and was occasionally fed by their father. Their mother tried to drive the young helper away; but once, while both rested on the nest's rim, this parent took food from the juvenile and passed it to a nestling. In an aviary, juvenile Indigo Buntings (*Passerina cyanea*) fed still younger birds, and Chipping Sparrows (*Spizella passerina*) about thirty-nine days old fed a younger Red-winged Blackbird (*Agelaius phoeniceus*).

Finches probably do not serve as interspecific helpers more frequently than birds of other families, but because they are among the most familiar birds in dooryards, gardens, and farms, their occasional assistance at nests of other

Northern Cardinals

species often comes to the attention of people who report it. Most of the records of helpers among finches that I have found belong to the interspecific category. Among them is that of a male Northern Cardinal who, after losing his own nest, turned his attention to four fledgling American Robins (*Turdus migratorius*). For a week he fed the young robins almost as frequently as their own parents did. Perfect concord prevailed among the three attendants. After the cardinals' replacement brood hatched, the male apparently brought food to both families alternately. His mate took no interest in the robins. Another male cardinal fed three nestling Yellow-breasted Chats (*Icteria virens*). Still another male cardinal fed seven goldfish, bringing food to them repeatedly for several days—standing at the pool's edge, he placed it in their open mouths raised slightly above the water. This strangest of all the cases of interspecific feeding that have come to my attention is corroborated by an excellent photograph.

Northern Cardinals receive as well as give help in interspecific transactions. A male Brown Towhee (*Pipilo fuscus*), whose own first brood had just become independent, joined a pair of cardinals in feeding three fledglings of the latter species. For about three weeks, the three adults worked together without the slightest discord. A month after helping feed the cardinals, the towhee and his mate reared a second brood. For hours together, a female Rufous-sided Towhee (*P. erythrophthalmus*) fed two young Northern Mockingbirds (*Mimus polyglottos*). The mockingbirds rejected seeds but accepted insects from their benefactor. A Rufous-sided Towhee and a Field Sparrow (*Spizella pusilla*) built nests only 18 inches (46 centimeters) apart in the same tree and hatched their eggs at about the same time. The male towhee frequently fed the young sparrows and removed their droppings, and a parent sparrow reciprocated by bringing food to the nestling towhees. Another Rufous-sided Towhee cared for a young Brown-headed Cowbird (*Molothrus ater*) that had been hatched in the nest of an Orchard Oriole (*Icterus spurius*).

Until their own young hatched, a pair of Song Sparrows helped a neighboring pair of American Robins feed their nestlings and clean their nest. One or two Song Sparrows helped Yellow Warblers (*Dendroica petechia*) nourish the warblers' brood. When the female warbler was on the nest, she took a sparrow's offering and gave it to a nestling. A pair of Song Sparrows and a pair of Northern Cardinals nested simultaneously in a nest built by the cardinals and lined by the female sparrow. Both females laid eggs and both incubated and brooded, with the cardinal sometimes sitting above the sparrow. Three cardinals hatched and were reared until they fledged, fed by all four adults cooperating closely without friction.

When a pair of Dark-eyed Juncos and a pair of Bewick's Wrens (*Thryomanes bewickii*) nested on opposite sides of the interior of a garage in California, the juncos often chased the wrens when they came with food for their nestlings. Despite this hostility, the male junco fed the wrens while his mate incubated, and he also cleaned their nest. The wrens did not try to drive the juncos away. A White-throated Sparrow (*Zonotrichia albicollis*) adopted a young Dark-eyed Junco. An unmated male of the vanishing dusky race of the Seaside Sparrow (*Ammodramus maritimus*) abandoned his territory in Florida to feed two recently fledged Red-winged Blackbirds. Their mother was completely tolerant of the helper, despite his efforts to keep her away. Attracted by the persistent calling of newly fledged Hawfinches (*Coccothraustes coccothraustes*) in England, a male Chaffinch (*Fringilla coelebs*) fed them six times.

In Colorado, a nest contained four half-grown American Robins, two newly hatched House Finches (*Carpodacus mexicanus*), and four eggs belonging to the latter. Its attendants were two adult robins and two female finches apparently mated to the same male, all five of whom fed the young regularly. The much larger robin nestlings smothered their nest mates. Despite the loss of their own progeny, the three finches continued to feed the robins after they fledged. At another nest in the same locality, adult robins and adult finches fed young robins. There was no evidence that the finches had laid eggs in this nest.

Occasionally one sees a bird with a bill so malformed, or otherwise handicapped, that it appears unable to nourish itself. The inference is precarious, for once I watched a Lineated Woodpecker (*Dryocopus lineatus*) feed himself with a bill so grotesquely misshapen that this seemed impossible, and similar cases have been reported for other birds. To be certain that the bird with an impediment is nourished by one or more companions, we must actually witness this. One such case was that of an adult male Black-headed Grosbeak (*Pheucticus melanocephalus*) with a deformed bill that evidently made it difficult for him to feed himself. He was fed by a female of his kind, by whose care he had apparently been kept in good condition for an extended period. A somewhat similar case was that of an ailing Black-faced Wood-swallow (*Artamus cinereus*) nourished by companions, as told in chapter 40. In a cage, a Eurasian Linnet (*Carduelis cannabina*) fed an injured adult of the same species. Also in captivity, male Eurasian Siskins (*C. spinus*) frequently fed adult subordinate males, although females were not seen to feed mature females.

REFERENCES: Allan 1979; Alsop 1973; Antevs 1947; Bell and Hornby 1969; Bent 1949; Brackbill 1944, 1952; Fox 1952; Greenlaw 1977; Hawksley and McCormack 1951; Hoyt 1948; Ivor 1944a; Jehl 1968; Lack 1953; Lemmons 1956; Logan 1951; Moermond 1981; Mountfort 1957; Neff 1945; Nice 1943; Nolan 1965; Price et al. 1983; Rakestraw and Baker 1981; Rice 1969; Senar 1984; Skutch 1954, 1961, 1972; Van Riper 1980; Westwood 1946; Williams 1942.

50.
WEAVERS AND SPARROWS
Ploceidae

Authors assign widely different limits to this family, some including all the Old World seedeaters, others defining it much more narrowly. Here I shall follow the latter course, which recognizes only three subfamilies, the buffalo-weavers (Bubalornithinae), the sparrows (Passerinae), and the true weavers (Ploceinae), and of these only the last two will be considered here. The ninety species of true weavers are confined to Africa south of the Sahara and neighboring islands, except for five in India and Malaysia. In size they range from that of a goldfinch (*Carduelis*) to that of a thrush (*Turdus*). Their bills are short and rather thick, and they eat both insects and seeds. Males of many species are brilliant with red or yellow and contrasting black; females are duller and sparrowlike. After the breeding season, the more ornate males molt into an eclipse plumage like that of females and young—a transformation rare among tropical passerines. With great skill, they weave elaborate covered nests with sideward- or downward-facing doorways or long entrance tubes. Some species breed in arboreal colonies conspicuous in the African landscape. Many are polygamous. The thirty-seven species of sparrows (not to be confused with the sparrows of the New World, which belong to a different family) are widely distributed over Eurasia and Africa and include the House Sparrow (*Passer do-*

mesticus), which we have helped to spread over nearly the whole Earth. Both subfamilies contain species with remarkable social habits.

Let us begin with five species of *Malimbus*, true weavers that live in dense tropical forests and have received less attention than those which inhabit savannas and other open country, where they are easier to find and to watch. These five weavers, studied by A. Brosset in Gabon in equatorial West Africa, exhibit ascending degrees of sociality which, however, fail to reach full cooperative breeding. The first is the Blue-billed Forest-Weaver (*M. nitens*), of which the male, working rapidly and alone, builds in two days a small globular nest with a downwardly directed entrance spout, hung from a vine, twig, or similar support at no great height above a stream or other water. When the nest is nearly finished, a female arrives. If she accepts the structure, she pulls off leaves and drops them around it, increasing its exposure, then lays two eggs. She alone incubates and broods the nestlings. The male guards the nest while she is absent and helps her feed the young with large larvae.

After a male Crested Forest-Weaver (*M. malimbicus*) chooses the nest site and starts to build, often beneath a frond of a rattan palm and frequently above water, he is joined by a female. The two work rapidly, and in about four hours they may finish their pensile nest, a crudely woven globular structure with a short, downwardly directed spout, much like that of the preceding species. Only the female incubates, while her mate keeps in touch with her by distinctive calls. Unfortunately, further details of nesting behavior are lacking, as none of the ten nests of which Brosset watched the construction survived long enough for him to learn how the young are attended. Some nests serve only as dormitories.

Rachel's Forest-Weaver (*M. racheliae*) frequents the upper levels of primary forest, where it is uncommon and rarely seen. One of its nests was built by two males and a female, working simultaneously.

Cassin's Forest-Weaver (*M. cassini*) attaches one of the finest examples of avian weaving beneath a palm frond at a good height. With narrow strips of palm fronds, interwoven in a crisscross pattern of great regularity, it fashions a yard-long nest that hangs like an inverted sock, of which the foot is the nest chamber and the leg a long entrance tube opening downward. So thin that it is transparent when viewed from the ground, the fabric is, nevertheless, so strong that it resists rain and wind for several seasons. Apparently, the female of this weaver chooses the nest site, starts to build, and attracts usually two males, less often one and rarely three. Whatever their number, they share the work of building with her, but she leads the undertaking and does as much as two males together. For a while, the three work amicably and are sometimes together in the nest with no indication of disagreement; but as the nest nears completion, the dominant male becomes aggressive and finally drives the helpers away, sometimes with the female participating in the chase. Even with three weavers, these more finely finished nests take about twelve days to complete. Of twenty-one nests, five were built by a pair only, thirteen by two males and one female, and three by three males and one female. After the departure of the helpers, the female lays her eggs and incubates them alternately with her mate, each sitting for about an hour at a stretch. Details of rearing the young are lacking. The birds sleep in their old nests.

Of the same form as the preceding nests and about 30 inches (76 centimeters) long, the nest of the Red-crowned Forest-Weaver (*M. coronatus*) hangs from the tip of a vine beneath the forest canopy, sometimes as high as 80 feet (24 meters). Of interwoven dry tendrils, twigs, and stalks, its exterior bristles with stiff, out-jutting stems, perhaps a device to discourage flightless predators. These curious structures are built by groups of adults of varying composition: two groups of one male and one female, two of two males and one female, three of one male and two females, four of two males and two females, six of three males and two females, and one of two males and three females. These builders never worked alone, and no leadership was apparent; all worked in exactly the same way. As in Cassin's Forest-Weaver, the dominant male appears to dismiss all but one of the other birds when the nest is finished, as only one male and one female shared incubating, brooding, and feeding the young. Thus, in contrast to the many polygamous weavers of open country, these five sylvan species are monogamous. Although their helpers were not permitted to participate in incubating and caring for the young, they gained valuable experience in building elaborate nests that would serve them when they in turn undertook to rear broods.

For more advanced stages of cooperative breeding, we turn from the true weavers to the sparrow-weavers, members of the weaver family that do not weave but build their covered nests by thatching. The White-browed Sparrow-Weaver (*Plocepasser mahali*), boldly patterned in white, black, and gray, is widely distributed over arid tropical Africa. It forages almost wholly on the ground, picking up fallen grass seeds or gathering them from growing grasses and searching for insects, sometimes rolling over stones, clods of earth, or other small objects with its bill or digging into the ground. In trees of no great height, it builds of grasses a domed nest about 12 inches (30 centimeters) long, entered through a spoutlike projection at one end. When one of these nests contains eggs or nestlings, it has only a single entrance, but when it serves as a dormitory it is open at both ends, an arrangement that permits the sleeper's escape through one doorway if a predator enters through the other. Often two nests, and sometimes three or four, are built in contact, foreshadowing the far-closer packing of nest chambers that we shall meet in the Sociable Weavers (*Philetairus socius*).

White-browed Sparrow-Weaver

White-browed Sparrow-Weavers were studied in Kenya by Nicholas and Elsie Collias and in Zambia by Dale Lewis. In both localities their social organization and behavior were essentially the same. They lived in groups of two to eleven birds on clustered, defended territories, which in Kenya were in acacia savanna and dry brush but in Zambia were in light woodland with a relatively closed canopy. Each group consisted of one monogamous breeding pair and their helpers, among whom a dominance hierarchy or linear peck order was evident, although dominance interactions were usually very rare among most members of a group. Each group had two or three times as many nests as members, in the same or neighboring trees. All members participated in building and maintaining nests throughout the year but were stimulated to greater constructive activity by the advent of the rains. However, they built with dry grasses, gathered from the ground, where they picked up loose straws or clipped them from standing plants. Another source of materials was their old nests, which were continually being dismantled. Nonbreeding adults slept singly in their nests with two doorways.

In Zambia, nesting continued for nine months of the year. Sets of one to three (average 1.4) eggs were incubated without help by the single breeding female of each group, who sat for intervals that averaged about eleven minutes, alternating with recesses of the same length, so that she covered her eggs only half of the daylight hours. After the young hatched, they were brooded only by their mother, who at first brought most of their food. After a few days, other group members started to feed the young, until most or all of them participated in this activity, bringing many small moths and caterpillars. Strangely, the dominant individual in each of two groups, presumably males, fed infrequently.

The one or less often two fledglings reared in a nest fluttered out when sixteen or seventeen days old. In the week before a young sparrow-weaver's departure, the adults of its group all joined in building for its accommodation a new nest with a conspicuous doorway to facilitate the fledgling's entry. As night approached, the young bird's mother and sometimes other group members would go in and out of the nest while the fledgling perched nearby, guiding and encouraging it to enter—as I have seen in wrens, araçaris, and other birds. Usually the young bird slept in the nest provided for it, but occasionally it entered some other nest. If this happened to be an adult's dormitory, the erstwhile occupant moved to some other nest, leaving the fledgling to sleep alone.

Nourished at first by all group members, the fledgling sparrow-weavers began to find some food for themselves about the second week after their departure from the nest. In a large group, the mother soon laid again, leaving the care of her young to the helpers. Juveniles continued to beg for food until they were over three months old and able to support themselves. During their first six months, they contributed little to nest building, but at nine months of age they became active nest helpers. One female was watched coaxing her older offspring to feed her latest brood. After both arrived together at the nest with food in their bills, the mother entered to feed the nestlings in the young bird's presence. Then she flitted back and forth between the youngster and the doorway until the latter followed her example and took the food inside.

In Zambia, the study area of 247 acres (1 square kilometer) supported about 108 White-browed Sparrow-Weavers in 26 groups of 2 to 11 individuals. These groups laid an average of 2.5 clutches per year. The total number of fledglings for the three years ranged from 0 to 12 per group. The mean numbers of fledglings per year for small, medium, and large groups were, respectively, 0.6, 1.2, and 2.3. Of 86 nests in which eggs were laid, 46 percent produced fledglings. Of 81 young fledged, 60 percent survived for the next six months; and of 35 six-month juveniles, 78 percent lived to be one year old. Neither the success of nests nor the survival of young during their first year was correlated with the size of groups, but the survival of reproductive females was strongly correlated with group size, being 0.0, 0.3, and 0.5 for small, medium, and large groups. Apparently, a breeding female survived better in larger groups because her assistants eased her parental burdens, especially by reducing her work load in nourishing the young; the more numerous her helpers, the less often she brought food to them, while the total number of meals received by nestlings increased with the number of their attendants.

A peculiarity of the White-browed Sparrow-Weaver's social system is that lost reproductive females are replaced by members of the same group, especially if it be large, whereas males are replaced by outsiders. In other cooperatively breeding birds, the reverse appears to be more frequent, with a younger male of the same group inheriting the place of the lost breeding male, and a female from a different group replacing the lost reproductive female.

Gray-capped, or Gray-headed, Social Weavers (*Pseudonigrita arnaudi*) were studied by Nicholas and Elsie Collias in the semiarid acacia savannas of Kenya. These birds lived in groups or families of three to seven individuals, whose activities centered about a number of nests in an acacia tree. A single small tree might support the nests of from one to five groups, each of which occupied a certain part of the branching crown; together they formed a colony of occasionally sixty or more individuals. The trees harboring different colonies were sometimes little more than 100 feet (30 meters) apart. Sometimes two to eight nests were built in contact. Throughout the year, the social weavers slept, singly or as many as five together, in domed nests of grasses, each with two downwardly directed doorways at opposite ends. Within a colony, a social hierarchy developed, but these birds rarely behaved aggressively toward members of their own group. They did not repulse from their nest sites members of other groups in the same colony, but they drove away visitors from a different colony. By day the social weavers spent most of their time gathering grass seeds and insects from the ground surrounding their nest tree. They did not defend a feeding territory, and at any season they might join members of other colonies in a large flock of foragers.

Gray-capped Social Weaver

As in other birds of arid regions, the breeding of Gray-capped Social Weavers is dependent upon rainfall. Each group had a single reproductive pair. Recently fledged young and older young from earlier broods helped their parents build, sometimes as many as six individuals working on one nest. When a nest was chosen for breeding, one of the two doorways was closed before the female laid her two or three eggs. For about fourteen days, the male and female incubated alternately, she taking sessions that averaged 8.5 minutes, while his sessions were less than a third as long, 2.5 minutes on the average. Sometimes both parents passed the night with the eggs, at other times only one.

During the twenty days that the young social weavers remained in the nest, they were fed by two to six attendants, including the two parents and immature as well as adult helpers of both sexes. At first the nestlings received mostly insects, but before they fledged grass seeds were added to their fare. The rate of feeding per nestling varied greatly, from 6.2 to 27 times per hour. The parents, working about equally hard, brought most of the nestlings' meals, but auxiliaries in adult plumage contributed substantially and immature helpers added their mite. When helpers were present, each nestling received an average of fifteen meals per hour, but at nests without helpers they received only ten. Young birds continued to be fed by parents and helpers for three or four weeks after they left the nest. Many remained in their natal area for at least nine months, probably furnishing the majority of their parents' assistants in the following breeding season.

To other advantages of cooperative breeding, some of which were mentioned at the end of our account of the White-browed Sparrow-Weaver, we may add the more rapid provisioning of the young after an interval of enforced neglect. Both the sparrow-weaver and the social weaver lose much time from foraging, and consequently from feeding nestlings or fledglings, when menaced by one of their many enemies, particularly Gabar Goshawks (*Melierax gabar*). When the peril has passed, the parents and their assistants resume feeding at an accelerated rate.

Among the few small birds that build nests conspicuous from afar are the Sociable Weavers. Their huge structures dot the barren landscape of the parched Kalahari Desert in Namibia and neighboring parts of southern Africa, monuments to the industry of these sparrow-sized birds clad in streaked brown and buff, with black throats. The preferred site of the Sociable Weavers' many-chambered nest is a stout Camelthorn Acacia (*Acacia giraffae*) with an open crown, but other trees with thick, more or less horizontal branches may be chosen, and the birds also build on telephone poles and the towers of water tanks. With sticks from 4 to 12 inches (10 to 30 centimeters) long, they arrange a dome-shaped roof or superstructure above the supporting limb. Inserting dry straws into this mass, where they are held by friction, the weavers extend the substructure downward below the branch. In this principal part of the nest mass they make the nesting chambers, each of which has a vertical entrance tube up to 10 inches (25 centimeters) long, leading upward to a cavity about 6 inches (15 centimeters) in diameter, set to one side, so that the whole compartment has somewhat the shape of a chemist's retort. Where the entrance tube joins the chamber, the birds construct a firm threshold of flexible grasses to keep the eggs from rolling out. They line the chamber all around with soft dry materials, including grass inflorescences, furry leaves, shredded grass blades, and a few feathers. These rooms do not intercommunicate. The entrance to each is surrounded by the ends of stiff, outwardly directed straws, which form a bristly collar that may repel a human hand and perhaps certain predators.

In warm, rainy regions, these nests of perishable vegetable materials would never resist decay long enough to attain their enormous size. But in arid and semiarid southern Africa, the birds add to the nest mass year after year until it may become huge. A nest in eastern Transvaal, belonging to a colony reported to be one hundred years old, was an irregular mass almost 16 feet long by about 12 feet wide (5 by 4 meters) and several feet thick. Its lower surface was pitted with the entrances of 125 nest chambers. Most nest masses are smaller than this, down to those with fewer than a dozen chambers. A colony, which may comprise up to five hundred birds, may start, on different branches of the same tree, several nest masses that grow and finally coalesce. Or the group may occupy two or more neighboring trees.

In an intensive study continued for

nineteen months in the Kalahari Gemsbok National Park, Gordon Maclean learned that all colony members join in starting a nest mass. When they are not breeding, three or four may collaborate on a single chamber, but after breeding begins each pair maintains its own. Like the more elaborate of the nests of ovenbirds, the Sociable Weavers' structures require unremitting attention to keep them in good repair. Throughout the year, the birds work at their nests whenever they are not resting, foraging, or attending eggs or young. They gather most of their materials, including pieces that have fallen from the nest mass, from the ground beneath or near it; but most birds returning from a foraging expedition bring back a straw, often from a point a mile away. Except the soft materials for the chamber's lining, which may be collected until the bird has a billful, a weaver carries a single piece, often a straw grasped by its lower end. When only eighty days old, Sociable Weavers begin to build. In their strong, constant impulse to build, Sociable Weavers resemble the ovenbirds that construct large nests of interlaced sticks.

As a nest mass grows, a curious division of the colony develops. Over part of the lower surface, the birds fill all the chambers with straws, and beneath them they build a new layer of chambers. By the repetition of this process, a large nest mass comes to have several levels. The occupants of each level participate in maintaining the roof that shelters all, but those of one level are repulsed if they intrude upon a different level. Inhabitants of the same level dwell in amity, with no apparent peck order or social hierarchy. All adults, however, are dominant over all immature birds until the latter acquire adult plumage, at the age of about four months, when they are accepted as full members of their group. Strangely, allopreening, so frequent among cooperatively breeding birds, is rarely practiced by Sociable Weavers.

The massive nests are the permanent abodes of their builders. Throughout the year, the birds sleep in them, each choosing whatever chamber it wishes in its own structural level. As many as five may occupy a single room, where they must sleep in layers. At dawn they begin to call in their nests, but usually they do not become active until sunrise or a few minutes later. Then in flocks they seek their feeding areas, usually within a mile of their nests. Here they spread over the ground, advancing by hops or a short, shuffling step or two, while they search for the insects that are the mainstay of their diet or gather the seeds of grasses. Around ten o'clock in the morning they return for a long siesta in their nest chambers, protected from the intense desert sunshine through the hottest hours of the day. At about two o'clock in the afternoon, they sally forth again for another spell of foraging. Around sunset, they flock back to retire into their nests. This, with certain variations, is their schedule in the warm days of summer. On a chilly winter morning they may delay in their snug nests much longer. But in winter their midday rest is often curtailed, and their evening return delayed until well after sunset, to compensate for the fewer daylight hours when they can forage.

Sociable Weavers have a varied vocabulary, but their song is not impressive, and they do not proclaim possession of a territory. Like many birds of arid regions, they have no definite annual period for breeding but await the showers that at any time of year awaken the dormant vegetation with its accompanying profusion of insect life. According to the duration of favorable conditions, their nesting may be brief or protracted, and in a prolonged drought they may fail to breed at all. A good shower is promptly

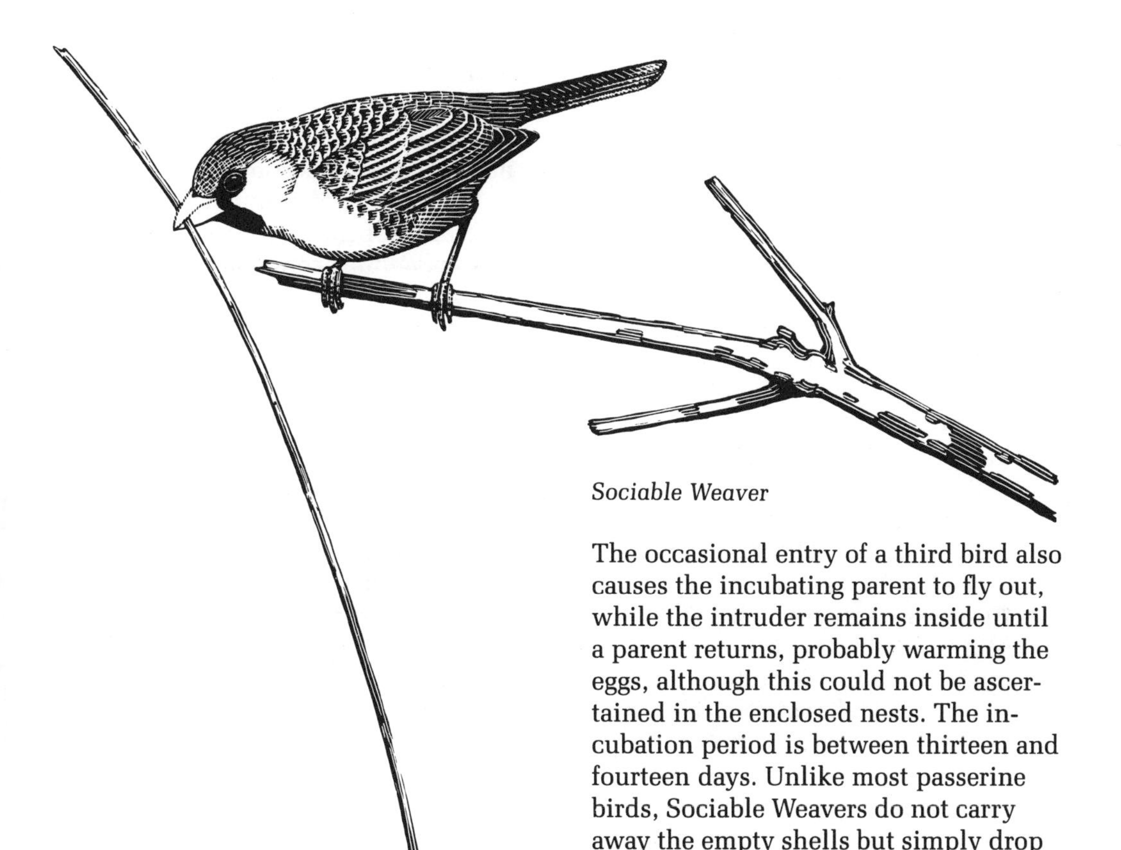

Sociable Weaver

The occasional entry of a third bird also causes the incubating parent to fly out, while the intruder remains inside until a parent returns, probably warming the eggs, although this could not be ascertained in the enclosed nests. The incubation period is between thirteen and fourteen days. Unlike most passerine birds, Sociable Weavers do not carry away the empty shells but simply drop them to the ground, where they hardly add to the conspicuousness of the nest mass or the accumulation of debris beneath it.

Typical passerine nestlings with sparse natal down, newly hatched Sociable Weavers have swollen, creamy white protuberances at the corners of their mouths that must make them conspicuous in the dimly lighted chambers. At first, the parents brood them by turns almost continuously. Both bring them insects that increase in size and number as the days pass. Harvester termites, which after a shower in the Kalahari Desert become active in large numbers by day, enter largely into the diets of nestlings of all ages. Although usually only the parents feed nestlings of first broods, occasionally one or two nonbreeding adults assist them. Another departure from the usual routine of passerine birds is the

followed by laying. The clutch usually consists of three or four eggs but ranges from two to six, being larger after a good shower than when conditions are less favorable. Chambers that serve as dormitories throughout the year are now used for rearing young, but many remain unoccupied by breeding pairs.

The male and female parents share incubation rather equally, sitting for two to forty minutes at a stretch and keeping their eggs almost constantly covered, as the incubating bird seldom leaves the nest before its mate arrives to replace it.

parents' failure to carry away the nestlings' droppings, which are not enclosed in fecal sacs to facilitate their removal. From the day they hatch, the nestlings void their wastes over the entrance tube, through which they fall to the ground.

After the fledgling weavers leave the nest, at the age of twenty-one to twenty-four days, they return to it to rest and sleep. They might not find the entrance to their own chamber among so many others if the parents, who recognize them individually amid the throng, did not guide them to it. The fledglings also recognize their parents, whom they approach for food with begging calls and soliciting postures. Although the young birds beg less intensely from all adults, only parents were seen to feed their fledglings, continuing this for about a week after their first departure.

In a favorable interval, Sociable Weavers may raise as many as four broods in nine months of almost continuous reproduction. As soon as the nestlings of the second brood hatch, young of the first brood, when only twenty-five to thirty days old, help their parents feed them. With each successive brood, the number of young helpers increases; a fourth brood was attended by eleven birds, including the parents and nine juveniles from three earlier broods. After acquiring adult plumage when about four months old, most young remain to breed in their natal colony.

Despite their well-enclosed nest chambers and the help they receive from juveniles, Sociable Weavers breed with very poor success. Of 2,789 eggs laid, 1,168 hatched, and 365 nestlings, representing 13.1 percent of the eggs laid, survived to fly from the nests. Most losses of nestlings were due to predation, chiefly by the Cape Cobra (*Naja nivea*), which can climb a tree and pillage all the nests of a small colony. Honey badgers also raided many nests.

In addition to breeding in avian apartment houses rather than in separate nests, Sociable Weavers differ in several respects from other cooperative breeders. Group participation in the building and maintenance of nest masses places them in the category of mutual helpers, and juvenile helpers are present, but other ways of helping common among cooperative breeders appear to be rare or absent. With a sex ratio in the area of Maclean's study of eight males to five females, one would expect unmated adult males to serve frequently as nest helpers. The sexes are indistinguishable, and some of the adults that occasionally entered nests with eggs or brought food to nestlings may have been males other than the father, but at best they appear to have contributed little.

The peculiar features of the weavers' breeding system may be attributed to their precocious maturity and short lives. Whereas advanced cooperative breeders such as Yellow-billed Shrikes (*Corvinella corvina*) and White-winged Choughs (*Corcorax melanorhamphus*) seem to require at least four to six years to reach reproductive maturity, on the available evidence Sociable Weavers rarely survive for more than three years and a few months, and they are capable of breeding at an early age. Moreover, in periods of prolonged drought, the weavers may starve in large numbers while they fail to breed and replace their losses. Their reproductive success is as low as that of small birds of tropical forests, such as White-bearded Manakins (*Manacus manacus*), with a much longer life expectancy. Sociable Weavers cannot afford to pass a year or more as nonbreeding helpers but, to replace their losses, must begin to reproduce at an early age and undertake as many broods as the climate permits.

Only two other, unrelated birds build apartment nests comparable to those of Sociable Weavers, but never attaining such great size. One is the plain brown, streak-breasted, starling-sized Palmchat (*Dulus dominicus*) of Hispaniola, the most conspicuously abundant bird on the island. Its large nest masses of interlaced sticks are usually placed in the royal palms that are a prominent feature of the landscape in many parts of the Dominican Republic, but other kinds of palms and branching dicotyledonous trees often support them. Most of these nests are high, from 20 to 75 feet (6 to 23 meters) above the ground. The largest that I saw, built completely around the crown shaft of a tall royal palm, was estimated to be 10 feet high by 4 feet in diameter (3 by 1.2 meters). I could not learn how many chambers it held or how many birds were attached to it, but at least twenty-five chats were flying to and from a much smaller nest in the same locality. In the evening, Palmchats, like Sociable Weavers, retire to sleep in their nests. They breed in compartments that do not intercommunicate, but the details of their reproductive system, and whether they cooperate in ways other than building their bulky nest masses, are questions that await elucidation. Likewise the Monk Parakeets (*Myiopsitta monachus*) of Argentina, Paraguay, and Uruguay, unique in a family that nearly always breeds in cavities in trees or niches in cliffs, construct great apartment nests of sticks; apparently these have not yet been thoroughly studied.

It might be supposed that nothing new remained to be learned about a bird so abundant, widespread, familiar, and frequently studied as the House Sparrow. Nevertheless, in a three-year investigation of this satellite of ours on and around the campus of Mississippi State University, James Sappington banded hundreds of birds and watched 254 nests, of which 161 (63.4 percent) were attended by helpers as well as the two parents. Nests with auxiliaries received an average of 254.1 daily visits, most or all to feed the young, as opposed to 206.6 visits to nests provisioned by the parents only. Two or three days before they left their nests, the nestlings became strangely quiet, lying crouched inside, and the helpers no longer fed them. Parents rarely brought food to the young birds on their last day in the nests, before they emerged. The helpers accounted for 12.4 percent of the feeding visits at those nests which they attended. In spite of this assistance, nests with helpers yielded no more fledglings than those attended by the parents alone. This statistical study fails to clarify the status of the helpers, their ages and relation to the parents, whether they were breeders who had lost their nests or nonbreeders, nor does it reveal how many assisted at a single nest. Much still remains to be learned about this ubiquitous bird.

Despite their reputation for aggressiveness, House Sparrows occasionally befriend birds of other species. In Kansas, a female sparrow fed three fledgling Eastern Kingbirds (*Tyrannus tyrannus*), whose parents were not seen. For at least a week, the sparrow continued to bring bread and other food to the flycatchers, not without a degree of hazard to herself, for her head was usually caught by the closing of the wide mouths into which she thrust her offerings, and to release herself she had to struggle. An almost identical episode was reported from Louisiana, where, also, no adult kingbird was seen to take an interest in the three fledglings whom the female sparrow fed. For ten days she continued to nourish at least one of the young kingbirds.

Another House Sparrow helped a pair of Red-eyed Vireos (*Vireo olivaceus*) feed and protect their nestlings. Still another female sparrow fed nestlings of the Tufted Titmouse (*Parus bicolor*) three times in forty minutes, in a nest box near that in which she was caring for her own brood. Other House Sparrows have fed nestling Cliff Swallows (*Hirundo pyrrhonota*), Tree Swallows (*Tachycineta bicolor*), and Yellow Warblers (*Dendroica petechia*). In Europe they have nourished Dunnocks (*Prunella modularis*), Eurasian Blackbirds (*Turdus merula*), Spotted Flycatchers (*Muscicapa striata*), and Serins (*Serinus canaria*). In an extensive survey of the literature, Marilyn Shy found ten cases of interspecific helping by House Sparrows, more than for any other species. Probably these sparrows are not more inclined to feed alien young than many other birds are, but in consequence of their wide distribution and close association with humans, more such incidents have been noticed and reported. It is not surprising that birds so responsive to the begging of alien nestlings and fledglings as House Sparrows are should frequently become nest helpers in crowded colonies of their own kind.

I have devoted a long chapter to weavers and sparrows because, more than any other avian family that has been well studied, they exhibit the whole range of cooperation among birds, from help in nest building only to assistance throughout the nesting cycle, the construction of large apartment nests, and interspecific helping.

REFERENCES: Brosset 1978; Collias and Collias 1964, 1978a, 1978b, 1980; Fitch 1949; G. D. Hamilton 1952; Hofman 1980; Lewis 1982a, 1982b; Maclean 1973; Prescott 1967; Sappington 1977; Shy 1982.

51.

THE SIGNIFICANCE OF INTERSPECIFIC HELPING

The foregoing chapters contain many instances of birds of one species feeding or otherwise attending young of different species. To these may be added a few examples of such behavior that did not belong in these chapters. In Alaska, a pair of Arctic Loons (*Gavia arctica*) adopted five ducklings of the Spectacled Eider (*Somateria fischeri*), hatched near the loons' nest on a small islet in a lake. The loons fed their fosterlings from their bills, as ducks do not, and carried them on their backs while they swam. The ducklings responded appropriately to the warning calls of their foster parents. A Screech Owl (*Otus asio*) and a pair of Northern Flickers nested in holes in the same tree. After the owl's eggs were destroyed, she was found on five consecutive days brooding the young flickers, while the latter, always uninjured, were regularly fed by their parents. The owl even brought a small bird to the young woodpeckers! In Mexico, a young White-eared Hummingbird (*Hylocharis leucotis*) was fed by a Green Violet-ear Hummingbird (*Colibri thalassinus*), a bird much heavier than a White-ear.

One of the first points that emerges from a survey of interspecific helping is the great disparity between the helpers and the helped in many of these episodes. Often they belong to different avian orders, with different diets or methods of feeding their young, as in the cases of the loons and the eider ducklings, the owl and the nestling flickers, the Common Starling and Northern House-Wren that also fed flickers, and the Mountain Chickadees who brought food to the Williamson's Sapsuckers. A bird that nests in the open may carry food to young in a box or hole, as the Gray Catbird did to Northern House-Wrens, or a hole-nesting bird may attend young in an open nest, as exemplified by the Blue Tit who fed nestling European Robins and the Eastern Bluebird who gave food to young Northern Mockingbirds. Likewise, birds may place food into nestling or fledgling mouths that differ in color from those of their own young. A Dark-eyed Junco, whose nestlings have red mouths, fed yellow-mouthed Bewick's Wrens, and a Red-legged Honeycreeper, also with red-mouthed nestlings, gave food to a fledgling Yellow-green Vireo that exposed a yellow gape. Or the mouth colors of helper and helped may be reversed, as when a Tropical Gnatcatcher, whose nestlings' mouths are yellow, attended the red-mouthed nestlings of Golden-masked Tanagers, and an Eastern Kingbird fed Baltimore Orioles. Strangest of all was the case of the Northern Cardinal who fed seven goldfish at the edge of their pond.

With few exceptions, all the instances of interspecific helpers that have come to my attention have been reported from the north temperate zone, in western Europe, the United States, and Canada. My only examples from the tropics are the Green Violet-ear who fed the White-eared Hummingbird and three that I have myself found, the Tropical Gnatcatcher and the Golden-masked Tanager, the male Red-legged Honeycreeper and the fledgling Scarlet-rumped Tanager, and the female Red-legged Honeycreeper and the fledgling Yellow-green Vireo. In contrast to this sparse harvest of only two species of interspecific helpers, I have watched intraspecific helping, habitual or occasional, in twenty-three species of birds in tropical America.

Why this great difference in the relative frequency of the two categories of helpers at low and at high latitudes? In the first place, intraspecific helpers, especially cooperative breeders, are, for reasons that will be given in chapter 53, actually far more numerous in the tropics and subtropics than farther north. But the paucity of records of interspecific helpers in tropical countries appears to be due largely to the fewness of those who are likely to observe and report them. Everywhere, interspecific helpers are more likely to attract the attention of people only casually interested in birds than are intraspecific helpers. To learn how many birds of the same kind are visiting a nest often requires concentrated watching of banded or otherwise individually distinguishable birds. But to notice, in the garden or outside the window, birds of different species bringing food to the same nest is so unusual that it arouses the curiosity of anyone even mildly interested in nature, who may report the occurrence to some professional or amateur ornithologist or to the biology department of the nearest university. Thereby, many of these occasions are carefully documented and finally published in an ornithological journal, the local birders' newsletter, or the daily press.

The records of interspecific helpers are so scattered that I cannot claim to have found more than a small fraction of them. I have none from South Africa, Australia, or New Zealand, where I doubt not that interspecific helpers have often been discovered. In the tropics, especially tropical America, people likely to notice and report unusual events in the bird world are much fewer and, until recently, most have been ornithologists from northern countries, often collecting or concentrating on particular studies and not likely to notice interspecific helpers.

Although, for the foregoing reasons, I believe that published records tend to exaggerate the frequency of interspecific helpers relative to intraspecific helpers, the former are certainly not negligibly rare. When I reflect upon the number and great diversity of helper-helped combinations included in the preceding chapters, the many reported cases that have undoubtedly escaped me, and the infinitesimal proportion of all bird nests that have been seen by humans, I suspect that almost every species has occasionally helped, and been helped by, every other species not too different in size and habits with which it has been associated over a large area for many years.

Although juveniles and older nonbreeding birds often help at nests of their own species, I have found only one mention of a juvenile interspecific helper among free birds—a young House Sparrow who fed nestling Tree Swallows, as reported by Marilyn Shy. In aviaries, however, juveniles supplied with abundant food frequently offer some of it to birds of different species with whom they are closely associated. Examples include a Black-shouldered Kite who adopted and reared a nestling buzzard, a six-week-old Eastern Bluebird who helped feed fifteen nestlings of several species, and the Japanese White-eyes who fostered nestling House Finches and House Sparrows.

Except in captivity, interspecific helpers are nearly always breeding adults. Male birds, and sometimes incubating females, are often so impatient to feed their expected nestlings that they bring food to the nest and present it to eggs far from hatching—the phenomenon that I have called anticipatory food bringing. If the female happens to be sitting when her mate arrives with food, she may or may not accept it, but often it is clearly not intended for her. Or a male bird may satisfy his premature impulse to feed by attending young in a neighboring nest, which among territorial birds will more often belong to a different than to the same species. Male Scarlet Tanagers of two different pairs fed nestling Chipping Sparrows while their mates incubated. Similarly, a male Northern House-Wren fed a brood of Northern Flickers in a hole near that in which his mate incubated, a male Dark-eyed Junco fed nestling Bewick's Wrens, and a male Eastern Bluebird attended nestling Northern House-Wrens. More rarely, both members of a pair with eggs feed a neighbor's nestlings, as happened when male and female Song Sparrows nourished a brood of American Robins. After their own young hatch, parents eager to begin feeding turn their attention to their offspring but may continue to bring more or less food to neighboring young.

Frequently, interspecific helpers are breeding adults who have lost their own brood or reared their young without exhausting their impulse to feed dependents. Included in this category are a male Northern Cardinal who, after losing his own nest, satisfied his thwarted parental instinct by feeding four fledgling American Robins. A male Brown Towhee whose first brood no longer needed care joined a pair of Northern Cardinals in feeding three fledglings of the latter. An Eastern Phoebe whose first brood was becoming independent transferred her attention to a brood of Tree Swallows. After rearing two of her own young, a female Eurasian Blackbird continued for two or three weeks to offer food to any bird who came near. Temporary thwarting of the impulse to feed offspring may divert this impulse to receptive young of a different species, as when a female American Redstart, whose young were being photographed in the hands of children, presented the contents of her bill to a neighboring brood of American Robins.

Sometimes parent birds become interspecific helpers by accident. When the collapse of a thin partition in a rotting trunk dropped a Mountain Chickadees' nest into a cavity occupied by nestling Williamson's Sapsuckers, the chickadees fed the young woodpeckers. As a parent Gray Wagtail (*Motacilla cinerea*) flew over a brood of young thrushes, they opened their mouths, whereupon the wagtail faltered in flight, turned, alighted, and gave all its food to them. A male Chaffinch, attracted by the persistent calling of newly fledged Hawfinches, fed them six times. A pair of Black-and-white Warblers fed a recently fledged Ovenbird who had apparently become separated from its parents. When two birds of rather similar habits have young of about the same age in nests close together, each is sometimes diverted to its neighbor's brood. Thus, a male Rufous-sided Towhee frequently fed nestling Field Sparrows, and the parents of the latter reciprocated by bringing food to the nestling towhees—a case of reciprocal altruism.

Recently the question of altruism in animals has been much discussed. If we regard altruism as a moral quality, dependent upon the state of mind or intention of the altruist, then we must withhold judgment, for the minds of animals are closed to us. If, on the other hand, we assess altruism solely by overt behavior, then interspecific helpers are undeniably altruists, for neither they nor their descendants derive any benefit from it. Their only reward is the satisfaction of their instincts or impulses—as is often true of our own uncalculating altruism.

The relations between helper and helped range all the way from harmony to mutual antagonism. Harmony prevailed between a helpful Northern Cardinal and parent American Robins, between the helpful Brown Towhee and a pair of cardinals, between a Northern House-Wren and parent Black-headed Grosbeaks, and between a Winter Wren and the Townsend's Solitaires at whose nest it assisted. Among helpers antagonistic to the parents were a Gray Catbird at a Northern House-Wren nest, one of the latter at a Northern Flicker nest, and a Tropical Gnatcatcher at a Golden-masked Tanager nest. Such antagonism sometimes increases as days pass and the helper develops a more proprietary interest in the nest where it assists. In other cases, the parents repel their uninvited assistant, as Worm-eating Warblers did to a helpful Black-and-white Warbler, and Tree Swallows did to a well-disposed Eastern Phoebe. Finally, helper and helped may be mutually hostile, each trying to keep the other away, as when a Common Starling fed nestling Purple Martins.

The Rufous-sided Towhee–Field Sparrow association mentioned above is an example of mutual interspecific helpers. Others are provided by birds of different species who share a nest. In a nest in which both laid, a Yellow-billed Cuckoo and a Mourning Dove incubated side by side. Another Mourning Dove shared with an American Robin a nest with two eggs laid by each, both of whom incubated and attended the nestlings. A robin and a House Finch took charge of a nest in which both had laid. A pair of Northern Cardinals and a pair of Song Sparrows tried to rear their broods in the same nest. These attempts to nest communally are likely to be partly successful only when the participants are more similar than doves and cuckoos or doves and thrushes; and at best only the young of the larger species are likely to survive until they fledge.

A survey of interspecific helping with all its strange combinations impresses one with the strength of birds' impulses to place food in other mouths, which overrides the limits of species, family, and order, disregards the appearance of the nestling or the color of the mouth into which the food is thrust, is not inhibited by a quite different nest site, persists in the face of strong opposition by the recipients' parents, and may occasionally impel a bird to neglect its own eggs while it feeds young of a different species. The impulsion to feed is one of the strongest in all the behavioral repertory of altricial, semialtricial, and subprecocial birds, because the perpetuation of their species depends upon depositing sufficient nourishment in the mouths of their young. It is one of the first components of parental behavior to become manifest in fledglings or even nestlings (although sometimes they make gestures of nest building at an equally callow age); it persists in birds unable to breed; it prompts bereaved parents to bring food to their nests for days after they have lost their nestlings; it may be redirected to social companions well able to nourish themselves; it enters into the courtship of many birds; applied to the bird's mate instead of the young, it helps form her eggs and sustain her while she incubates.

Even nest parasites that have lost most or all other parental behavior may retain the impulse to feed: the shining cuckoos (*Chrysococcyx*) of Africa have repeatedly been seen to feed nestlings or fledglings of their own species raised by foster parents, as well as to give food to their mates; and, more rarely, Brown-headed Cowbirds have been reported to feed young cowbirds. Finally, the impulse to feed may find expression in fantastic situations, as when a captive Common Raven passed food through the bars of its cage to a free Black Vulture, a Northern Cardinal fed goldfish in a pond, and a Jackdaw, regarding Dr. Konrad Lorenz as his mate, pushed food into the distinguished ethologist's mouth and even his ear!

The variety of nestlings and fledglings, and indeed animals of other kinds, which birds occasionally feed is proof that this behavior is not precisely adjusted to the features of the birds' own offspring. Apparently, instances of interspecific helping, although numerous in the aggregate, are too infrequent to diminish appreciably the reproduction of any species, and, accordingly, natural selection has failed to produce a finer adjustment. However, in another context, that of nest parasites and their hosts, greater discrimination by parent birds could save many species a huge drain on their reproductive output—a drain that has threatened the extermination of such rare species as the Kirtland's Warbler (*Dendroica kirtlandii*).

The parasitic female's chief problem seems to be to have her eggs accepted by the prospective foster parent, who may throw them out, cover them with nest materials, or abandon the violated nest. If the parasite's dupe accepts and hatches the alien eggs, it apparently never fails to feed the fosterlings, no matter how greatly they differ in size and appearance from its own young. Eggs of numerous Old World cuckoos (although not of New World cowbirds) frequently match those of the host closely in color and size, but mimicry of the host's young by the parasite's young is rare. It occurs in the parasitic whydahs, whose gaping nestlings expose the same curiously patterned mouths as their closely related weaverfinch (Estrildidae) hosts; in the Koel (*Eudynamys scolopacea*) and Great Crested Cuckoo (*Clamator glandarius*), whose black fledglings resemble those of their black corvine hosts; and in the Screaming Cowbird of South America, whose young are confusingly similar to their foster parent, the nonparasitic Bay-winged Cowbird. Evolution's failure to give frequent hosts of nest parasites the protection from parasitism that finer discrimination of the objects of their parental care might bring them is another indication of the irrepressible strength of their impulse to feed and to brood. If mating were not more strictly controlled than feeding, by more numerous and salient specific characters, avian hybrids would be more common than we find them.

Interspecific helping is evidence that birds' parental behavior occasionally escapes the strict genetic control which, if sufficiently strengthened by evolution, might confine their ministrations to their own offspring. The behavior, such as feeding young of a different species, is innate, but the context in which it occurs is obviously not narrowly limited by heredity, except in a few species such as the Sooty Tern (*Sterna fuscata*), which, if given eggs of the Brown Noddy (*Anous stolidus*) to hatch, will throw from the nest or kill the nestlings so similar to its own. The bird who attends alien nestlings appears to enjoy a measure of freedom, of self-determination or spontaneity—as it also does when it selects the situation of its nest amid a variety of potential natural sites. Release from the strict determination of activity by the genes is a most important component of freedom, especially our own: we owe almost everything that civilization has achieved, intellectual and moral as well as material, to our ability to think and act in ways not programmed in our genes. Interspecific helping reveals that at least some birds are sometimes free to innovate behavior, even to act capriciously, in ways that might have important consequences for their descendants.

Because interspecific helping is too sporadic to influence to a measurable degree the reproductive potential or the demography of any species, its significance has been largely neglected by students of the evolution and ecology of cooperative breeding. As far as we know, no avian species has been less successful in maintaining its numerical strength because some of its members occasionally spend some of their time and energy helping other species. Nevertheless, by revealing the great strength of birds' impulse to feed as well as their freedom to act in ways not strictly determined by their heredity, the study of interspecific helping may contribute to our understanding of the more advanced forms of intraspecific helping.

Although, evolutionists stoutly maintain, genetically controlled behavior always tends to maximize the number of an individual's own genes in its descendants, the same need not be true of behavior that escapes strict genetic control. Birds are occasionally free to behave in ways that they find satisfying, regardless of any effects upon reproduction. They may remain in their natal territory in a subordinate role, instead of hazardously emigrating to rear broods of their own, because they enjoy companionship and feel more secure in the home of their childhood. They may return to that home after failure to establish themselves elsewhere. They may build covered nests for more comfortable sleeping even if this takes time and energy that might be applied to reproduction. If the innovation jeopardizes the perpetuation of the species, it will be suppressed by natural selection. But if it is compatible with the continued prosperity of the species, it may persist in innovative individuals until it is supported by the process variously known as organic selection or genetic assimilation, whereby behavior that is at first individual and spontaneous is finally supported by mutations that establish it firmly in the heredity of the species.

To recognize that this is one of the ways in which cooperative breeding might arise is to attribute to the birds' choices, to their minds, far-reaching consequences in the evolution of avian behavior. Evolution can promote the development of an organ or function only if that organ or that function influences the course of evolution. The eyes of birds, for example, have been perfected because visual acuity has so powerfully influenced the evolution of birds. It is not otherwise with mind, which can modify behavior by discrimination, choice, and intelligence. If mind did not influence the course of evolution, as it obviously has done in our own species, it would never have developed beyond its first rudiments. Although birds' minds are much simpler than ours, they are neither negligible nor devoid of flexibility. The phenomenon of interspecific helping is evidence that birds enjoy a measure of freedom from strict control by their genes, and the exercise of this freedom may in the long run lead to advanced cooperative breeding.

References: Abraham 1978; Armstrong 1947; Bent 1938; Friedmann 1960, 1968; Lashley 1938; Shy 1982; Wagner 1959.

Note: In this and the following chapters, I give scientific names and references only when they are not given in the appropriate preceding chapters.

52.

CHARACTERISTICS OF COOPERATIVE BREEDERS

In the foregoing chapters, we broadly surveyed the incidence and activities of nest helpers and cooperative breeders among the families of birds. In the present chapter, we shall consider some of the widespread features of these cooperative birds. We shall concentrate our attention upon advanced cooperative breeders, whose young remain with their parents at least into the nesting season following that in which they hatched and are integrated into fairly stable social groups. Sometimes it is difficult to draw the line between advanced cooperative breeders and those somewhat less than advanced; but species that have only occasional helpers are excluded from the present discussion, as are those, such as Long-tailed Tits, whose helpers are chiefly breeding adults who have lost their young. Those with helpers of different species were considered in the preceding chapter.

Where do we find advanced cooperative breeders? Since long-distance migration or wide wandering in the inclement season causes the disintegration of families, except those of certain larger birds, such as swans and geese, cooperative breeding occurs mainly in the tropics and subtropics, where birds are continuously resident. Nomadism appears to be less disruptive of families than long-distance migration, and in Australia a substantial minority of cooperative breeders are nomadic. The Acorn

Woodpecker, one of the cooperative breeders that ranges farthest from the equator, up to northwestern Oregon at 45 degrees north latitude, can remain on its territory throughout the cold months thanks to its abundant stores of acorns and the snug holes in which it sleeps.

The number of cooperative breeders that have been carefully studied amid the dry, open woodlands, savannas, and arid scrub of Africa, Australia, and other regions with light vegetation might lead one to conclude that these are the habitats most favorable to this life-style. It would be truer to say that these are the habitats most favorable for studying cooperative breeding. Nests are relatively easy to find and to watch and, as in many of the drier parts of Africa, living accommodations are available to visiting scientists and it is easy to move about in a Land Rover. Cooperative breeders are found in almost every terrestrial and freshwater habitat in Earth's warmer regions that will support them, from arid deserts, such as the Kalahari and the Arabian, to the wettest rain forests. In two breeding seasons at what is now a field station of the Organization for Tropical Studies at La Selva, in the rain forest of the Caribbean lowlands of northern Costa Rica, we found seven species of cooperative breeders, three of which were not previously known to have this habit. But despite much searching, we discovered only one to three nests of each of these species, and the risk of losing these precious nests by too-frequent or too-close approaches limited what we could learn about them. Certainly many more cooperative breeders remain to be discovered in the rich avifaunas of tropical rain forests.

Just as cooperative breeders are not confined to any single habitat, so they are not restricted to a particular diet. Insectivores (in the broad sense) are in the majority, doubtless because insectivorous birds tend to be sedentary and strongly territorial, whereas many tropical frugivorous and nectarivorous birds wander widely as, now here, now there, a tree or shrub blossoms or ripens its fruits. Frugivorous and nectarivorous birds tend to the opposite extreme from cooperative breeding: to nesting by solitary females unassisted by males, who vie for their attention in leks or courtship assemblies, as in manakins, birds of paradise, and hummingbirds. But cooperative breeders include the largely frugivorous barbets, Collared Araçari, and Purple-throated Fruitcrow, the mainly vegetarian Acorn Woodpecker and Purple Gallinule, seed-eating weavers, and omnivorous jays.

Most cooperative breeders reside throughout the year on defended territories. They may be either facultative or obligate. The latter, which include the White-winged Chough, Yellow-billed Shrike, and Gray-breasted Jay, appear rarely if ever to nest successfully as unaided pairs. Among the former, which are far more numerous, many, often most, pairs lack helpers, frequently because they are birds who are breeding for the first time or who have lost all their progeny of earlier years, with the result that they have none to assist them. In addition to unaided pairs, cooperative breeders live in groups of three to about ten or twelve individuals, with extremes of twenty reported for the White-winged Chough, Jungle Babbler, and Gray-breasted Jay, twenty-five for the Yellow-billed Shrike, twenty-six for the San Blas Jay, and thirty-five for the Chestnut-bellied Starling. Some of these very large groups contain several breeding pairs. The huge nests of Sociable Weavers may have as many as five hundred inhabitants, who cooperate in maintaining the superstructure, while other modes of cooperation are confined to much smaller subgroups or families. In certain cooperative species, including the Southern Ground Hornbill, some individuals, mostly females, remain alone or with a companion or two of the same sex during the breeding season.

Advanced cooperative breeders tend to lack colorful plumage; often they are black, brown, or gray. Notable exceptions are the handsome Purple Gallinule, Green Jay, Tufted Jay, Golden-breasted Starling, and Superb Starling. Whether dull or brilliant, the sexes tend to be alike; even the male and female Red-cockaded Woodpecker are difficult to distinguish in the field, although the sexes of most woodpeckers are readily separated by their head patterns. Outstanding exceptions to this rule are the blue wrens of Australia, among which the elegance of the males contrasts strikingly with the plainness of the females. Apparently, subdued coloration and the absence of conspicuous sexual differences help reduce sexual jealousy among closely associated adults.

In contrast to the similarity of fully adult individuals of the two sexes, younger birds tend to retain for a long while features that distinguish them from their elders. These lingering signs of immaturity often involve the colors of bills, eyes, or bare facial skin. The black bills of fledgling Green Wood-hoopoes take a year or more to become red, as in adults. Bills of cooperatively breeding jays darken from yellow to black as the birds mature, a change that may take more than a year and that proceeds irregularly, producing variously pied patterns which facilitate the recognition of individuals. Until about six years old, Southern Ground Hornbills are distinguished by the colors of their eyes and the bare skin of their faces and throats. The irises of White-winged Choughs, brown at first, take about four years to become wholly red. Similarly, Gray-crowned Babblers' eyes brighten from dark brown to yellow in about two and a half years. Birds, whose plumage may change when they molt, appear to recognize their companions mainly by their heads. The maintenance of group cohesion and harmony appears to depend upon a subtle balance of likeness and difference, equality and subordination; small rather than striking differences in the appearance of a group's members seem to make this balance easier to preserve.

A group of cooperative breeders usually consists of a single breeding pair with one or more auxiliaries who are usually their offspring from an earlier year, often with one or more immigrants from other groups. Birds of the immediately preceding season, yearlings, often predominate among the helpers, but older individuals are frequent. Group members are usually, perhaps always, arranged in a hierarchy or rank order; however, this may not be obvious, because a peck order—such as develops among domestic chickens and certain other birds—is absent, and aggressive interactions may be rare. They are seldom so rough as the bill wrestling with wing blows by which older Laughing Kookaburras keep younger ones in their place. To learn the hierarchical order among gentler birds, investigators sometimes intensify competition to an unnatural level by offering a concentrated source of food, such as peanuts in a can.

In a typical rank order, the breeding male takes precedence over all other individuals in his group, who rank below him in descending order of their ages, the youngest at the bottom. The breeding female heads individuals of her sex, also ranked according to their ages. The breeding female may be subordinate not only to her mate but likewise to helpers subordinate to him. The persisting signs of immaturity that we have just noticed may help preserve this order. In most cooperative breeders, the hierarchy is not a system of exploitation or bullying but a gently maintained order of precedence that promotes concord and facilitates cooperation. It is hardly different from the situation in many harmonious human families, in which unmarried sons and daughters live with their parents and help maintain the household, and the father, or in matriarchal societies the mother, is head of the family and chief decision maker, and the children exercise authority in the order of their ages, experience, and strength of character.

In many cooperative breeders males outnumber females, often by as much as 1.5 to 1 or even 3 to 1. Exceptional are Pukekos, Laughing Kookaburras, White-winged Choughs, Gray-crowned Babblers, and Yellow-billed Shrikes, whose groups contain approximately equal numbers of the two sexes. Since males and females tend to be equally represented among fledglings, why do the former predominate in the adult population? The reason is that in most cooperative breeders females emigrate more often than males, who much more frequently remain in their natal groups. Wanderers confront more perils, and suffer higher mortality, than birds who stay on the familiar home territory, with the result that the sex more inclined to emigrate becomes less numerous. Another cause of the higher mortality of females is the fact that the sex which incubates, especially at night, suffers higher predation than other group members.

An exception to the greater tendency of females to emigrate is found in the White-browed Sparrow-Weaver, in which males wander abroad more often, and a lost reproductive female is replaced by a member of the same group. In other species, the oldest surviving male of the group usually inherits the rank of a lost breeding male, while the female breeder is more often replaced from outside. Sometimes both sexes emigrate, in parties of two to four of the same sex, who together seek another group or try to form a new group, in which they have better prospects of rising to the top. If they succeed, the oldest will become the breeder, and his or her companions will serve as helpers.

Among social mammals, females commonly stay in their natal group, while young males seek their fortunes elsewhere. Why female birds leave home more often than males is obscure. It can hardly be because they are subordinate to males, because this will be true in whatever group they enter. If they find a group that has lost its reproductive female, they may attain breeding status sooner than they would in their natal group; but apparently they are more likely to perish than to become breeders, resulting in the unbalanced sex ratio that we have noticed. Instead of remaining in comparative safety at home, females seem to sacrifice themselves to avoid inbreeding and to spread their group's genes more widely through the population, with possible benefit to their species. However, in some species they rarely go much farther than neighboring groups, and if their quest of a vacancy that they can fill proves futile, they may return after absences of days or weeks and be accepted by their erstwhile companions, as has been recorded of Florida Scrub Jays, Straight-crested Helmet-Shrikes, White-browed Sparrow-Weavers, and Black Tits and is probably generally true. Cooperative breeders are home-loving birds.

Most cooperative groups dwell in harmony, with rarely a serious dispute or aggressive behavior. Fledglings and juveniles of Jungle Babblers, Arabian Babblers, and probably others are often quarrelsome or unruly, but as they mature they learn better manners. Puerto Rican Todies, whose helpers have not been associated with them since the preceding breeding season but appear after their nesting is well advanced, first repel but later accept these volunteers. Instead of preserving an individual distance while they perch, many cooperative breeders are contact birds, who rest by day and sleep by night pressed against each other. Those that breed in open nests roost at night in a compact row on the branch of a tree or shrub or sometimes even more closely massed together, some on the backs of others. Species that raise their families in a roofed nest or a cavity of some sort commonly sleep together in the same or a similar closed space, so that even if they do not perch in contact by day, they must be packed together at night, in layers if they are numerous. Those that possess such dormitories usually lead their fledglings to sleep in them with their elders—a habit not confined to cooperative breeders.

Exceptional is the White-browed Sparrow-Weaver, in which group members build special dormitories for the fledglings, while adults sleep alone in separate lodges. Also unexpected is the behavior of Red-cockaded Woodpeckers, who leave their fledglings outside while each adult seeks its own particular hole for sleeping. This is the way of some other woodpeckers, including the Red-crowned (*Melanerpes rubricapillus*), in which helpers have not been found; but Golden-naped Woodpeckers, who have only rare, mostly inefficient juvenile helpers, lodge in a hole with their young until the following breeding season.

Among both birds and mammals, mutual grooming or preening ranks high among the ties that bind social animals together. While they rest in intimate contact, cooperatively breeding birds frequently preen each other. Allopreening, as this is called, occurs among Pukekos, Tasmanian Native Hens, anis, Green Wood-hoopoes, Southern Ground Hornbills, Purple-throated Fruitcrows, White-bearded Flycatchers, Green Jays, Yucatán Jays, Apostlebirds, sittellas, Banded-backed Wrens, babblers, blue wrens, wood-swallows, white-eyes, and others. It appears to be rare in Sociable Weavers, absent in Gray-breasted Jays, and unrecorded in some other cooperative breeders. Allopreening is certainly not peculiar to cooperative breeders, but in other birds it tends to be restricted to members of a pair, especially just before and during the breeding season.

Somewhat less frequent is exchanging food or feeding fully self-supporting companions, other than a female preparing to lay or engaged in incubating or brooding; but it occurs in Pukekos, Harris' Hawks, Pied Kingfishers, Yucatán Jays, Black Tits, babblers, wood-swallows, and Northwestern Crows. Although the attitude of the recipient, often with fluttering wings and begging cries, is often interpreted as a gesture of submission to the donor, it is probably no more than the persistence of the fledgling's habit of receiving its meals in this fashion. Northwestern Crows, Black-faced Wood-swallows, White-breasted Wood-swallows, and Laughing Kookaburras (and even several big birds that are not cooperative breeders) have substantially prolonged the life of an injured companion, unable to forage, by frequent gifts of food. From all these activities emerges a picture of groups of birds who, far from living in a state of tension and rivalry, dwell in relaxed friendship.

In a number of cooperative breeders, group members join in choruses or displays or both simultaneously. Among the nonpasserines, choruses tend to be loud and stirring rather than melodious. Such are the performances of White-fronted Nunbirds perching in a row high in a tropical rain forest, the far-carrying dawn chorus of Southern Ground Hornbills, and the startling outbursts for which the Laughing Kookaburra is named. At daybreak, Noisy Miners sing together for many minutes. Among the notable displays are the flag waving and rally of Green Wood-hoopoes and the huddles of Gray-crowned Babblers, all accompanied by much sound. Confrontations of two groups of cooperative breeders at their common boundary generally excite far more displaying than fighting. San Blas Jays and Gray-breasted Jays are reluctant to cross territorial boundaries.

Cooperative breeders defend their companions as well as their nests and fledglings. When a Gray-crowned Babbler is caught by a predator or held in a human hand, all babblers within hearing, including members of neighboring groups, rush toward it and join in a simultaneous distraction display. White-winged Choughs crowd around and defend a companion attacked by an aggressive Black-backed Magpie. White-breasted Wood-swallows vigorously attack predatory animals, flying or flightless. Collared Araçaris pursued a hawk that had caught a fledgling. Finally, cooperative breeders, young and old, are sometimes playful. The social frolics of Southern Ground Hornbills, Apostlebirds, Common Babblers, and Brown Jays are described in the appropriate chapters.

Most groups of cooperative breeders have a single breeding pair, composed of the ranking male and female, and a single nest at any one time. Occasionally, Acorn Woodpeckers, White-winged Choughs, and some populations of Gray-crowned Babblers nest communally, with two or rarely more females laying their eggs in the same nest. In promiscuous groups of Pukekos, two or three females frequently share a nest. Two or more pairs of monogamous anis often occupy the same nest and raise their young together. Instead of laying several sets of eggs in one nest, other cooperative breeders build a number of contemporaneous nests in their territory, each belonging to a breeding pair. Large groups of Common Babblers sometimes had two nests at the same time, but the second always failed. A group of Yucatán Jays may have two or three breeding pairs, Gray-breasted Jays up to six, and Southern San Blas Jays six to ten, all with as many separate nests. A large group of Chestnut-bellied Starlings has from two to six nests at one time. A group of Guira Cuckoos may have several nests, or several females may lay in the same nest.

All this looks much like colonial nesting, in colonies defended against other colonies by all members. However, when a cooperatively breeding group has several simultaneous or temporally overlapping nests, usually all of them are attended by many or all members of the group, including those with nests of their own, so that they are no less communal breeders with mutual helpers than the anis, to which this designation has long been applied. By reciprocally feeding each other's young, wood-swallows also become mutual helpers.

Several groups of Gray-capped Social Weavers occupy different parts of the same tree, forming a colony of groups, which they defend from intrusions by members of other colonies. Pied Kingfishers sometimes dig their burrows in loose colonies in inviting banks. Aerial flycatchers, including swifts, swallows, and bee-eaters, often nest in colonies. Since it is hardly possible to delimit territories high in the air whence most of their food is procured, their territories, if they may be said to have them, are little more than their nest sites. The nonmigratory Red-throated and White-fronted bee-eaters are advanced cooperative breeders. Despite the various ways in which migratory House Martins help their neighbors, they can hardly be admitted to this category because their families do not remain intact from one season to the next, and who helps whom appears to be largely a matter of fortuitous contacts. The same appears to be true of those other long-distance migrants, Barn Swallows and Chimney Swifts.

The number of helpers at a nest is always somewhat less than the number of birds in a breeding group because the parents are, by definition, not helpers (except among communal breeders), and other group members, especially juveniles, may not aid. From the single helper at some nests of nearly all cooperative breeders, their number ranges up to a recorded maximum of twelve at nests of Southern San Blas Jays, thirteen at nests of Gray-breasted Jays and Chestnut-bellied Starlings, and fourteen at nests of Bushy-crested Jays. These communal breeders with a number of contemporaneous nests in a territory tend to have more helpers at each nest than do species with only a single breeding pair; but once ten attended a nest of Green Wood-hoopoes, and nine have been identified at nests of Common Babblers and Sociable Weavers.

When their parents raise two or more broods in a season, juveniles, and even birds hardly past the fledgling stage, often serve as helpers. These young assistants may begin at a surprisingly early age—twenty-five or thirty days in Sociable Weavers, forty-eight days in Smooth-billed Anis, sixty days in Splendid Wrens, sixty-nine days in Purple Gallinules, seventy days in Brown Tree-creepers—and quite frequently when about three months old. In many cooperative breeders that seldom or never raise second broods, the majority of the auxiliaries are yearlings, hatched in the immediately preceding breeding season. From this peak the number of assistants declines as individuals emigrate, graduate to the position of breeders in their own group, or die. Two-year-old helpers are present in many species; and among Florida Scrub Jays, White-winged Choughs, Yellow-billed Shrikes, and Splendid Wrens, individuals five or six years old remain among the auxiliaries. The fact that a bird serves as a helper is no proof that it is sexually immature or that it is not old enough to breed if it could become a dominant member of a group. Laughing Kookaburras, Red-throated Bee-eaters, Chimney Swifts, Acorn Woodpeckers, Superb Blue Wrens, and probably many others are sexually mature when about one year old but at this age often serve as nonbreeding auxiliaries.

The activities (other than self-maintenance) in which helpers engage are, in the order of increasing frequency, incubating and brooding nestlings, nest building, feeding the incubating parent on or off the nest, feeding the young and cleaning their nest, and defending eggs and young—which is the order in which male birds of all kinds serve at their mates' nests. Helpers have been found incubating among Chimney Swifts, Puerto Rican Todies, Laughing Kookaburras, Red-throated Bee-eaters, Red-cockaded Woodpeckers, Acorn Woodpeckers, White-winged Choughs, and Common Babblers, in all of which the male parent shares incubation with his mate. Nearly always a bird who incubates performs the very similar activity of brooding. In species of which the breeding male does not incubate, helpers rarely do. As a rule, when juveniles participate in an activity, they are less diligent and efficient than yearlings and older birds. But in nest defense, especially against human intruders, youngsters often protest more vehemently and take greater risks than older birds, probably because youth tends to be reckless and imprudent.

Despite the number of attendants, some cooperative breeders nest with poor success, but they compensate for low fecundity by living long. Southern Ground Hornbills produced an average of one fledgling per group every 6.3

years. Their average life span was calculated to be about twenty-eight years. The much smaller Acorn Woodpeckers raised to the age of nest leaving only 0.2 or 0.3 young per adult per year, which is one fledgling for every five or every three adults in the population. Adult males, the more numerous sex, had an estimated life span of thirteen years, females one of about ten years. White-winged Choughs, who produced only 1.14 fledglings per nest, took four or more years to become adult, which suggests that they live a long while. Another bird whose delayed maturity indicates longevity is the Yellow-billed Shrike. After studying this bird for six years, Llewellyn Grimes had not learned at what age it is ready to breed; two six-year-old female auxiliaries had not yet nested. At the other extreme, little Sociable Weavers acquire adult plumage when about four months old, breed early, nest with very poor success, and appear rarely to survive for more than three years. They compensate for their high losses of eggs and young and their short life expectancy by nesting three or four times in years when frequent showers help keep the desert green.

53.

BENEFITS AND EVOLUTION OF COOPERATIVE BREEDING

A pattern of life so widespread as cooperative breeding, adopted by so many species of birds of different families in such diverse habitats throughout Earth's warmer regions, must certainly give some substantial benefits to its participants. What are the benefits and who are the beneficiaries? This question may be divided into two: what does cooperative breeding contribute to the welfare of individuals, and how does it affect the reproduction and stability of the species?

Let us look first at the young birds who serve as helpers. Instead of the hard lot of all those inexperienced juveniles who are cast upon their own resources in a perilous world soon after they can feed themselves, the young of cooperative breeders remain with their parents on the home territory, guided by adults and shielded by them from many dangers. Although, like most young birds, they are innately equipped to support themselves, their inherited capacities can be strengthened and improved by experience. From their elders they become familiar with the most productive spots for foraging and the safest refuges from danger; by participating in mobbing, they learn to recognize enemies. As in the flocks of mixed species that wander through the woodlands, so in cooperative groups many keen eyes and alert ears bring prompt warning of the approach of a predator that might pounce unperceived upon a solitary individual.

By sleeping with companions in a snug dormitory or even by pressing close to them on an open perch, the young bird conserves heat and energy on cool nights. Its associates may preen and remove vermin from parts of its plumage inaccessible to its own bill. The individual bird doubtless feels safer in a group than by itself; that it actually is safer is attested by the fact that the sex which stays at home, usually the male, survives much better than the sex that more often emigrates—leading to the unbalanced sex ratios that we have noticed in many cooperative breeders. Male Galápagos Hawks prolong their lives by becoming members of nesting trios or quartets instead of remaining in the floating nonbreeding population.

Moreover, by helping with its parents' subsequent broods, the young bird acquires experience that will increase its efficiency when it in turn undertakes to rear a family. In earlier chapters, we saw how Yellow-tailed Thornbills and Brown Jays became more proficient builders of nests or feeders of young by assisting their elders. We have a growing body of evidence, chiefly from birds that are not cooperative breeders, that experienced birds breed more successfully than novices. The educative experience might be gained more cheaply by assisting at another's nest than by trying inexpertly at one's own. A helper often becomes a breeder by inheriting the rank of its deceased father or mother or by joining some other group. In the former situation, the nestlings that it has fed, now grown, may reward it by serving at its nest.

For the parents, the gains from cooperative breeding are equally great. The auxiliaries commonly help defend the territory, which may become larger or of better quality thanks to their assistance. Often they help build a nest, especially a dormitory where the whole family sleeps warm and dry. Some helpers take turns at incubation, thereby giving the parents more time for self-maintenance, or they feed the incubating parent. Usually they contribute substantially to feeding nestlings and fledglings, lightening a burden which for unaided parents with large broods can be debilitating. Some of the results of such relief were noticed in earlier chapters.

In double-brooded species, such as the Laughing Kookaburra and the Superb Blue Wren, parents with helpers are much more likely to have the energy to nest again in the same season, and to do so successfully, than are unaided parents. While the parents are engaged with the later nest, their assistants care for the fledglings of the first brood. Common Babbler parents with three or more helpers could molt—a process costly in terms of vital resources—while they had nestlings, as those with fewer assistants rarely did. Breeding female White-browed Sparrow-Weavers survived best in large groups, as did subordinate parents at the communal nests of Groove-billed Anis, where the dominant male alone undertook the dangerous duty of nocturnal incubation. Another advantage of having helpers is that the young could be more rapidly fed after a prolonged interval of enforced neglect while a predator was in view, as frequently happened to groups of White-browed Sparrow-Weavers and Gray-capped Social Weavers. James Counsilman concluded that scarcity of food made helpers necessary to feed young Gray-crowned Babblers and abundant predators made them needed to defend the nest.

Some students of cooperative breeding have surmised that helpers are manipulated or exploited by the parents; others, that they are spongers who impose upon their parents. Neither is probable: if auxiliaries feel that they are overworked or harshly treated, they are free to leave; if the parents find their helpers burdensome or annoying, they are able to expel them from the territory. Others ask whether helpers are altruistic or selfish. If by "altruistic" we mean benefiting others with no reward beyond the satisfaction that this beneficence brings, and by "selfish" we mean exploiting or manipulating others for one's own gain, neither of these attributes is applicable to avian auxiliaries. Helpers are neither more nor less altruistic or selfish than the boy who, unpaid, aids his father in the chores of a farm or the girl who lightens her mother's household tasks. Each gives and receives benefits without calculating the amount. Just as a wholesome child helps its parents without remembering that it may inherit from them, so the helpful bird aids its elders unaware that it may rise to breeding status on the parental territory, however much this possibility may have influenced the evolution of cooperative breeding; to believe otherwise is to ascribe longer foresight to young birds than to human children. The relationship of breeding adults and their auxiliaries in cooperative breeding systems is just what the term implies: cooperation or mutualism. Sharing of efforts and rewards binds individuals into a coherent society; pure altruism is often a unilateral relationship that emphasizes the inequality of individuals; slaves are compulsory altruists.

When evolutionists or sociobiologists ask whether nest helpers are altruists, they usually give to this term a special meaning, closely associated with the concept of fitness. When we hear of "the survival of the fittest," we assume that the fittest are the strongest, healthiest, most capable animals, best able to find food and escape enemies. The modern evolutionist has a single measure of fitness: the number of progeny an organism produces in its lifetime. At first sight, this appears an arbitrary restriction of the word's meaning; but we may reflect that in a state of nature, the strongest, healthiest, most capable animals are the ones most likely to beget the greatest number of equally capable offspring. An altruist, in the special meaning of the term, is an individual that sacrifices its own fitness—the number of its progeny—while augmenting the fitness of another.

Whether helpers are altruists in this special sense is a question difficult to answer. Possibly those who serve as auxiliaries for a single breeding season are not; although they contribute to the fitness of the parents whom they assist, they may be preparing themselves to breed so much more efficiently, with less risk to themselves, that they compensate for the loss of one season when they might be attending their own nests. However, to give a definite answer to this question, we would need to weigh the probability of winning a territory and a mate by emigrating against the probability of inheriting the territory on which they serve. The relatively few birds who continue as helpers for three, four, or five years do indeed appear to be altruists; but if the probability of finding a territory and a mate is very slight, perhaps they do the best they can for themselves. I would leave the question of whether intraspecific helpers are altruists in this special sense to mathematical sociobiologists, with their assumptions and complicated calculations, and continue to regard these auxiliaries simply as collaborators in the business of perpetuating their species and their way of living. Most obviously altruistic are the interspecific helpers.

This brings us to one of the theories of the evolution of cooperative breeding, which ascribes it to habitat saturation.

Some avian species have an optimum habitat, where breeding is possible, adjoining or at no great distance from a marginal habitat, where individuals can survive but insufficient amounts of food, cover, or nest sites prevent reproduction. Excess individuals overflow into the marginal habitat, awaiting an opportunity to enter the optimum habitat and breed if and when vacancies occur. They provide a standby population that may help replenish a breeding population somehow depleted. But for some birds in some regions, marginal habitat is sparse or lacking because of the absence of an ecological gradient, or intermediate zone, between areas where breeding is possible and those where individuals can hardly survive. Young birds who cannot find an opening in the optimum habitat must either remain on the parental territory or perish. Not surprisingly, they prefer the first of these alternatives. Some who try fruitlessly to raise their status elsewhere return to their natal group.

The theory briefly outlined implies that cooperative breeders are successful species that fill their habitat to its carrying capacity. It does not ascribe their success to the presence of helpers but, at least, since these species continue to saturate their habitat, the auxiliaries are no detriment. Like the marginal populations of other birds, the helpers are a reserve to replenish losses among the breeding adults or to take advantage of such opportunities for range expansion as might arise. Moreover, they are a superior reserve, for they already have experience in nest attendance, which a marginal population of nonbreeders lacks. Cooperative breeding is rightly viewed as a flexible method of population regulation. Ornithologists in Australia, where cooperative breeders are numerous, view this system as especially appropriate for the unpredictable climate of their mostly arid continent. In drier years when food is hard to find, the helpers may be needed for provisioning the young; when more generous rainfall creates more favorable conditions, the auxiliaries may help rear second or third broods or, already experienced, occupy newly tenable areas.

Theorizing that habitat saturation promotes cooperative breeding seems to apply to birds like Acorn Woodpeckers in California, if not over their more extensive tropical range. Their special requirements of acorns, trees for storing them, and trees for drilling sap wells in the bark are not everywhere to be found. The theory does not appear applicable to the Banded-backed Wren, one of the most adaptable birds that I know, at home in open woods of pine and oak high in the altitudinal temperate zone, at the edge of tropical rain forests, and in shady seaside gardens. Nor does it seem to fit the Brown Jay, an aggressive bird which for many avian generations has been occupying new territory as the Caribbean rain forests of Middle America, the interior of which it avoids, have receded before clearings with scattered trees, where it thrives. Evidently, diverse roads lead to cooperative breeding.

An avian species that has filled all available space, as the habitat saturation theory implies, and that is, moreover, long-lived might advantageously restrain its reproduction, thereby reducing the expenditure of valuable resources to rear progeny whose prospects of survival are slight. Here we meet the knotty problem of the regulation of reproduction, a subject beset with paradoxes, contradictions, and disagreements among biologists. An influential school, of which David Lack was a leading protagonist, holds that every organism, birds included, must rear as many sound progeny as it can because, if it does not, its lineage will be submerged by more prolific members of its species. This seems difficult to reconcile with a dominant trend in the evolution of animals, which for ages has been to produce fewer offspring and take better care of them, as is evident when one compares the great number of eggs laid and neglected by fish and amphibians with the small broods or litters, often of only one or two, produced and carefully nurtured by birds and the more advanced mammals.

Moreover, excessive reproduction may be self-defeating. For many animals—even for birds in evergreen tropical rain forests—seasons of more abundant food, when they breed, alternate with seasons of less abundance, when reproduction is suspended. If animals overload their habitat in the favorable season, many will succumb in the leaner months that follow, but before dying they will have consumed food that might have kept others alive until the following season of greater abundance, so that species which produce more than a certain optimum number of offspring may enter the next breeding season with smaller populations than they might have had if they had not reared so many young. This optimum rate of reproduction might be defined as that which can be exceeded only at the price of a mortality that will cancel all its gains—which, indeed, might have disastrous consequences in the form of widespread disease or starvation. Although maximum reproduction might give natural selection more material on which to operate, the price, in wasted resources and lives prematurely extinguished, is very high.

Birds avoid excessive reproduction by various means. In the humid tropics, where annual mortality is low, many species lay only two eggs in their nests; some lay only one. Some birds devote to the time-consuming construction of elaborate nests energy that they might apply to rearing larger broods. Males of many species remain totally aloof from nests while they indulge in visual and vocal displays to attract females, thereby halving the labor force available for rearing the young. Many birds both terrestrial and marine delay breeding for several years after they have grown to adult size, in sharp contrast to what happens in many mammals. Whether or not some of these birds could increase their reproductive output as evolutionary orthodoxy requires, by laying more eggs, building simpler nests, or breeding at an earlier age, is a question endlessly debated by ornithologists.

Among the birds that delay breeding are the helpers in cooperative breeding associations. In most of the foregoing accounts, we noticed that pairs with helpers produce more fledglings per nest, or rear more of their young to independence, than do simple pairs of the same species; if they did not, they would confirm the doubts of those who question whether helpers really help. In some species, perhaps most, the auxiliaries' contribution to breeding success is greater between nest leaving and independence than between egg laying and fledging; but since the latter period is easier to quantify, and nesting success is usually expressed as the percentage of nests or of eggs that yield fledglings, how much the helpers increase the survival of the young after they leave the nest is not revealed even by some of the most prolonged studies. Although in general helpers increase the success of nests and the survival of mobile young, exceptions are not lacking. Among the Pukekos studied in New Zealand by John Craig, simple pairs raised two or three times as many young per family, and three to six times as many per adult, as larger groups did. Many factors contribute to breeding success: the simple pairs may have been more experienced birds or may have occupied areas with more abundant food or fewer predators. Sandra Vehrencamp's study of Groove-billed Anis in Costa Rica showed that the advantage of having more nest attendants was greater in some habitats than in others.

Does cooperative breeding realize the maximum reproductive potential of the birds engaged in it? This is an important question because if it does not, these birds do not rear as many young as they might do if all nested as simple pairs, and cooperative breeding might be included among the means that birds use to adjust their populations to their resources. This would accord well with the theory that relates this breeding system to habitat saturation, but it would clash with the theory of maximum reproduction. To answer this question, we must examine the output of young per individual attendant rather than per nest or per group.

From his pioneer study of Superb Blue Wrens in eastern Australia, Ian Rowley concluded that groups with helpers annually produced 1.9 fledglings per adult, whereas simple pairs raised only 1.2. However, after a more prolonged investigation of Splendid Wrens in western Australia, the same author wrote: "Whether fledglings or yearlings are considered, the seasonal production by simple pairs does not differ from that of groups with helpers." More frequently, groups with helpers rear more young per nest or per season than unaided pairs, but the latter produce more young per adult, as in the Harris' Hawk, Galápagos Hawk, Red-throated Bee-eater, and Puerto Rican Tody. In other species, including the Groove-billed Ani, Acorn Woodpecker, Florida Scrub Jay, White-winged Chough, Jungle Babbler, and Common Babbler, the number of young per adult rises with one or a few helpers but remains constant or falls after they exceed a certain number—the parents have more auxiliaries than they need for efficient reproduction. Although the evidence is not wholly consistent, it points to the conclusion that cooperatively breeding birds do not raise as many fledglings as they might if they nested as simple pairs or with fewer assistants, but they compensate for reduced fecundity by the better survival of fledged young and the greater life expectancy of adults, which may be due, among other reasons, to the fact that cooperative breeding lightens the burden of parenthood.

In this book our chief interest has rested in how cooperative breeding affects the welfare of the helpful birds themselves, giving them safer, longer, apparently more satisfying lives. Evolutionary biologists have been mainly interested in what happens to the genes that control the birds' heredity, of which the birds themselves are unaware. Adult birds who devote their time to rearing others' offspring instead of multiplying their own genes in their own progeny perplex evolutionists, because they are difficult to reconcile with current evolutionary theory. To overcome the difficulty, biologists have recourse to the concept of kin selection. A large proportion of helpers, but by no means all, assist their own parents to rear younger brothers and sisters. Siblings have many genes in common. Since the chromosomes that bear the genes are transmitted to offspring much as cards in shuffled decks are dealt to players, in extreme, improbable cases siblings might have identical sets of genes (without reference to identical twins, which are not known to occur in birds), or all their genes might be different. On the average, siblings should share half their genes, which is often expressed by saying that their relatedness is one-half. Cousins and more distant relations have correspondingly fewer identical genes.

Accordingly, the helper who attends its younger brothers and sisters, or even its cousins, helps multiply many genes just like its own, including those that make it a cooperative breeder. Indeed, if a nest with auxiliaries is likely to be much more productive than the nest of an unaided pair, the helper may contribute more to the multiplication of its genes by attending it as a nonbreeder than by breeding in a solitary pair. And when its chances of acquiring a territory and a mate, even of surviving, apart from its natal group are slight, it will certainly transmit to posterity more genes identical with its own by helping at home than by venturing abroad. Its inclusive fitness (the number of genes identical with its own that it helps transmit to future generations through its brothers and sisters or more distant relatives, as well as through its own children if it has any) may be high even if its individual fitness is low or zero.

Kin selection fits cooperative breeding into the general evolutionary structure that biologists have so painstakingly erected, but it is not the whole story. It hardly accounts for the many intraspecific helpers who come from other groups and who may not be related to the parents whom they assist; all that is needed for a bird to become a helper is its acceptance by a group, as Jerram Brown's membership hypothesis recognizes. This tells us how cooperative breeding can persist even in the many species which each year have a substantial proportion of pairs nesting without auxiliaries, but it hardly even suggests how cooperative breeding arises. For this I offer the following theory, based upon two widespread avian traits.

Birds cling to the known and shrink from the unknown. If one has been long in a cage, it may be in no hurry to escape, even when the door is open. As I have repeatedly seen, juveniles resist expulsion from the family territory where they grew up or exclusion from the family dormitory. Migratory birds appear to be an exception, for without external compulsion they undertake long, perilous journeys. I suspect that migrants, especially those that travel by night, fly in the mental state of somnambulists, who, it is said, walk safely in their sleep where they would fear to tread while awake. I doubt that the small nocturnal migrant is aware of the marvelous navigational feat that it accomplishes when it alternates each year between two pinpoints on the map, perhaps thousands of miles apart—a feat that a trained human navigator accomplishes with charts and a whole panel of instruments. I believe that the bird is guided subconsciously by a wonderful innate mechanism, perfected by countless generations of natural selection and hardly understood by ornithologists. The nocturnal migrant does not hesitate or resist separation from its natal spot because it does not know what it is about to do. Moreover, if it fails to migrate it will probably starve or freeze in the inclement season.

The situation is different with the small, nonmigratory bird of warm lands. Its parents, having reared it to independence, try to expel it from their territory, so that they can breed again without interference and with less revealing movement around their nest. The juvenile, attached to the only place that it knows, tries to stay. If it belongs to a species that rather fully occupies its optimum habitat and has little marginal habitat, so that to leave home may be suicidal, it will cling more stubbornly to its birthplace, not because it anticipates what will befall it if it leaves, but because natural selection will have strengthened its innate determination to remain. The same selection pressure will have mitigated the parents' drive to expel their self-supporting progeny. Even a juvenile of a species not so limited by habitat will often resist exile into the unknown. If it clings tenaciously enough to its birthplace, or if the parents are indulgent, it may be permitted to stay. Then, closely associated with its siblings of the following brood and fairly competent at foraging, it will be driven by the bird's strong impulse to place food into begging nestling mouths, which we noticed in chapter 51, to feed its younger brothers and sisters. It will become a juvenile or a yearling helper.

After remaining on the parental territory for a year or more, the young bird may spontaneously leave; but now it is more experienced, better able to take care of itself, than a juvenile barely self-supporting. Often it emigrates with siblings of the same sex, and frequently it goes no farther than a neighboring territory which it already knows and in which it has detected a vacancy. Its departure from the home territory is much less hazardous than that of a juvenile unfamiliar with the wider world; and if it fails to find another domicile, it can return home and be accepted.

For this situation to become stable and hereditary, two changes are necessary. The parents' impulse to drive away their young must weaken. The young helper must remain subordinate to its parent of the same sex, and it must remain sexually quiescent even if it becomes reproductively mature, lest disruptive conflicts arise. It will be recalled that waywardness of the young female Southern House-Wren, as told in chapter 32, led to prolonged fighting and the final expulsion of the too-precocious helper. If the continued presence of grown young in the parental domain increases the survival of nestlings, fledglings, helpers, or the parents themselves, the necessary modifications will probably evolve. Persisting traces of immaturity—as in the colors of bills, eyes, or facial skin, as we noticed in chapter 52—will help maintain the group's rank order by indicating the status of its younger members.

Chapter 51 contains abundant evidence that birds frequently behave in ways not strictly determined by their heredity, although everything that an animal does must have an innate foundation. The first step in the evolution of a cooperative breeding system may well be a young bird's spontaneous refusal to leave its natal territory. If such home-loving youngsters frequently remain and contribute to the welfare of their own close relations, kin selection should promote the process of genetic assimilation whereby, through supporting mutations, beneficial behavior that was originally individual and spontaneous becomes inherent and widespread in the species. By such alterations, the group of parents and their grown-up offspring becomes a stable, harmoniously cooperating association.

By their reluctance to abandon their natal spot and their readiness to respond to any gaping nestling mouth, birds seem preadapted for cooperative breeding, which has evolved independently in many families in the most varied habitats. Since the behavioral traits at the root of cooperative breeding are widespread in birds, we may ask why this system is not more common than we find it. One reason appears to be that the increased activity at a nest may draw the attention of predators. Antbirds (Formicariidae), eminently adapted to life in the lower levels of predator-ridden tropical American forests, build inconspicuous nests and minimize their visits to them by taking long incubation sessions and bringing their two nestlings infrequent, large meals. Since so many nests are prematurely lost, small broods appear more appropriate for this situation than large ones attended by more birds and more expensive to replace. A greater number of attendants may be able to repulse certain less aggressive predators that the unaided parents could not successfully confront, but their presence might not substantially increase the security of a nest whose safety depends chiefly upon remaining undetected.

In other cases, the territory might not yield enough food to support adults additional to the parents, who without aid must exert themselves strenuously to provision their young. Although helpers usually increase the yield of nestlings per nest, they do not always augment the number per individual adult, and some species may need the greater reproductive potential that early breeding by unaided parents might give them. This is especially true of birds that suffer high mortality on long migrations, which, moreover, disrupt the family bonds that are the basis of cooperative breeding. For a variety of reasons, many avian species may not be able to afford the luxury of cooperative breeding, even if it would be more congenial to them.

Communal nesting, in which two or more females lay in a nest attended by all of them and frequently also by their mate or mates, is sometimes practiced by species with nonbreeding helpers, including anis, Acorn Woodpeckers, Pukekos, White-winged Choughs, and Gray-crowned Babblers. Incubation and brooding by three or more parents taking turns greatly reduce the burdens and the risks incurred by each participant, all of which would seem to make this breeding system attractive to birds. Why is it not more widely practiced? The difficulty of synchronizing laying by the several females weighs against it. The nestlings of the female who lays last will usually hatch last, be smaller, and fail to receive their due share of food in a crowded nest; some or all may starve. To avoid this, the last female ani or Acorn Woodpecker removes earlier eggs before she starts to lay. The other females continue to deposit eggs to compensate for their losses but may end with smaller sets than the last female. This results in more synchronized hatching and a more equitable production of fledglings by the associated females, but it is a crude, wasteful method of overcoming the difficulty. Often, too, more eggs than can be properly incubated are laid in a communal nest, and the activity there makes it too conspicuous. Moreover, females tend to be less tolerant of female coworkers than males are tolerant of males. For all these reasons, only a fraction of nests are communal even in species that most often practice this system.

The other variety of communal nesting, with several pairs occupying as many separate nests in a group's territory and all or most reciprocally helping each other feed the young, as in Chestnut-bellied Starlings and several species of jays, is also infrequent but for different reasons. Its rarity suggests that special, little-understood conditions are necessary for this system to originate and to prosper.

The retention of young by their parents is not the only road to cooperative breeding. Sometimes adults unite in a nesting association, as when two male Galápagos Hawks or Harris' Hawks mate with a single female and join in attending her nest. Disparity in the numbers of the sexes is likely to promote such arrangements.

For cooperative breeding to arise and persist in any avian species, it need not produce more, or as many, fledged young per nest or per adult as nesting by unaided pairs might yield, but by prolonging the average life span of all cooperating individuals, it should be at least as successful in maintaining a flourishing population as the alternative of earlier breeding by individuals in simple pairs that is still an option of many cooperatively breeding species.

Our attention is first drawn to a pattern of life exhibited by only a minority of avian species—and hardly known half a century ago—by noticing that more than two birds are bringing food to a nest or otherwise attending it. Then, if we are professional ornithologists, we try to demonstrate, by prolonged study and by gathering a mass of numerical data, that this rather exceptional breeding system is compatible with individual selection, or kin selection, or group selection, or whatever interpretation of evolution the investigator favors. We should be more open to the possibility that the three modes of selection operate together. Many of the birds in cooperatively breeding groups are, or become, parents, often after serving for a while as helpers. According to their competence and that of their assistants, as well as their success in escaping the hazards that afflict all nesting birds, they produce more or fewer progeny. Many of the individuals also help rear siblings, again with varying degrees of productivity. Moreover, groups differ in the number of progeny that they rear and the number that emigrate to spread their genes among neighboring groups. Accordingly, at each of these levels—the individual, the sibling, and the comprehensive group—we find the differential survival and reproduction that natural selection implies. It is obvious that individual selection, kin selection, and group selection are simultaneously involved in cooperative breeding.

Prolix discussions as to which of these modes of selection is predominant tend to divert our attention from the wider significance of this life-style, the enduring family bonds, the enhanced cooperation, the rise of life to higher levels of social integration and harmony. Group members do not exchange food, preen one another, build dormitories and sleep in contact, join in group displays, and engage in similar intimacies because the nonbreeders among them help the breeding pair raise the latest brood; on the contrary, they become helpers because they are already so closely associated with the parents and their nest. The intimate association throughout the year is primary, helping at the nest for a fraction of the year one of its consequences—just as exposure to nestlings may induce unrelated birds of the same species, or even those of different species, to attend them.

Unfortunately, the widely used designation "cooperative breeding" tends to distort the situation by emphasizing one aspect of a far more comprehensive association. "Family unity" would more adequately describe this development, the highest expression of family life among birds, if not among vertebrate animals except, perhaps, the most harmoniously integrated and cooperative of human families. Although "cooperative breeding" is inadequate as a designation of the most closely integrated groups, it is useful because it covers a wider spectrum, including species in which occasional helpers are less intimately associated with the parents.

Family unity is of varying degrees and takes diverse forms even among closely related species. Golden-naped Woodpeckers sleep with their parents in the family dormitory until nearly a year old; but as juveniles Golden-napes contribute little or nothing to the care of a second brood, and they are not known to help as yearlings. By contrast, a Red-cockaded Woodpecker will not share a hole with another individual past the nestling stage, even when holes, which take a long while to carve, are so scarce that members of its family must roost clinging to the outside of a trunk or fly afar to surreptitiously enter a vacant cavity in the territory of a different family. Social integration and the benefits it brings to individuals may amply compensate for a concomitant reduction of fecundity, especially when the reproduction of the cooperative breeders is adequate to maintain the species in a flourishing state, and additional young would have poor prospects of surviving.

In contemporary evolutionary theory, animals are viewed as little more than machines for producing the maximum number of offspring, regardless of the quality of their lives. This may be true of more primitive organisms, but more advanced animals may resist the tyranny of reproduction and numbers while they seek a more gratifying existence. Birds are not feelingless mechanisms for the multiplication of their genes but sentient creatures concerned for their own safety and comfort. I surmise that many birds breed cooperatively because they value the feeling of security that comrades give, enjoy companionship, and find this a satisfying way of life. These, of course, are values that are not amenable to scientific investigation or to mathematical treatment by evolutionary biologists, but we cannot for that reason dismiss them as fanciful or unimportant. To the philosopher or indeed to anyone seeking to fathom life's meaning, they are of paramount interest.

If, despite its crude gambling methods and many miscarriages, evolution did not raise at least some of its creations to higher levels of awareness and enjoyment, it would be a prolonged, complicated, harsh futility, an endless sequence of transformations producing nothing of worth; and it could make no slightest difference to any creature whether it continued or ceased to exist. All of which reminds us once more of the pathetic limitations of our sciences, which leave almost totally unexplored beyond the narrow human realm the psychic aspect of the Universe, for this aspect, which is probably coextensive with the material aspect and appears to be intensified as beings rise in the scale of creation, gives to existence all the value that we can discover in it.

References: In accordance with the plan and purpose of this book, in this chapter I have dealt rather summarily with the theoretical aspects of cooperative breeding. For those who might wish to pursue the subject in depth, the following titles from a very extensive literature should be helpful: J. L. Brown 1974, 1978, 1984; Emlen 1978; Lack 1968; Ricklefs 1975; Skutch 1967, 1976; West-Eberhard 1975; Woolfenden 1976; Zahavi 1976.

BIBLIOGRAPHY

Abraham, K. F. 1978. Adoption of Spectacled Eider ducklings by Arctic Loons. *Condor* 80: 339–340.

Acker, J. R. 1977. Premature parental behavior of a Red-shouldered Hawk. *Auk* 94: 374–375.

Addicott, A. B. 1938. Behavior of the Bush-tit in the breeding season. *Condor* 40: 49–63.

Allan, T. A. 1979. Parental behavior of a replacement male Dark-eyed Junco. *Auk* 96: 630–631.

Allen, A. A. 1930. *The book of bird life*. New York: D. Van Nostrand.

Alley, R., and H. Boyd. 1950. Parent-young recognition in the Coot *Fulica atra*. *Ibis* 92: 46–51.

Alsop, F. J., III. 1973. Notes on the Hoary Redpoll in its central Canadian Arctic breeding grounds. *Wilson Bull.* 85: 484–485.

Alvarez, H. 1976. The social system of the Green Jay in Colombia. *Living Bird* 14 (for 1975): 5–44.

Anderson, A. H., and A. Anderson. 1962. Life history of the Cactus Wren. Part 5: Fledging to independence. *Condor* 64: 199–212.

Andrews, M. I., and R. M. Naik. 1965. Some observations on flocks of the Jungle Babbler *Turdoides striatus* (Dumont) during winter. *Indian J. Ornith.* 3: 47–54.

Antevs, A. 1947. Towhee helps cardinals feed their fledglings. *Condor* 49: 209.

Armstrong, E. A. 1947. *Bird display and behaviour: An introduction to the study of bird psychology.* London: Lindsay Drummond.

———. 1955. *The Wren*. London: Collins.
Balda, R. P., and J. H. Balda. 1978. The care of young Piñon Jays (*Gymnorhinus cyanocephalus*) and their integration into the flock. *J. für Ornith*. 119: 146–171.
Balda, R. P., and G. C. Bateman. 1971. Flocking and annual cycle of the Piñon Jay, *Gymnorhinus cyanocephalus*. *Condor* 73: 287–302.
——— and ———. 1973. The breeding biology of the Piñon Jay. *Living Bird* 11 (for 1972): 5–42.
Baldwin, M. 1974. Studies of the Apostle Bird at Inverell. Part 1: General behaviour. *Sunbird* 5: 77–88.
Beason, R. C., and L. L. Trout. 1984. Co-operative breeding in the Bobolink. *Wilson Bull*. 96: 709–710.
Beebe, W. 1918. *Jungle peace*. New York: Henry Holt.
Bell, B. D., and R. J. Hornby. 1969. Polygamy and nest sharing in the Reed Bunting. *Ibis* 111: 402–405.
Bell, H. L. 1982. Social organization and feeding of the Rufous Babbler, *Pomatostomus isidori*. *Emu* 82: 7–11.
———. 1983. Co-operative breeding by the White-browed Scrub-Wren *Sericornis frontalis*. Emu 82, suppl.: 315–316.
Bellrose, F., Jr. 1943. Two Wood Ducks incubating in the same nesting box. *Auk* 60: 446–447.
Bent, A. C. 1938. Life histories of North American birds of prey, part 2. *U.S. Natl. Mus. Bull*. 170.
———. 1940. Life histories of North American cuckoos, goatsuckers, hummingbirds and their allies. *U.S. Natl. Mus. Bull*. 176.
———. 1942. Life histories of North American flycatchers, larks, swallows and their allies. *U.S. Natl. Mus. Bull*. 179.
———. 1946. Life histories of North American jays, crows, and titmice. *U.S. Natl. Mus. Bull*. 191.
———. 1948. Life histories of North American nuthatches, wrens, thrashers and their allies. *U.S. Natl. Mus. Bull*. 195.
———. 1949. Life histories of North American thrushes, kinglets, and their allies. *U.S. Natl. Mus. Bull*. 196.
———. 1953. Life histories of North American wood warblers. *U.S. Natl. Mus. Bull*. 203.
———. 1958. Life histories of North American blackbirds, orioles, tanagers, and allies. *U.S. Natl. Mus. Bull*. 211.
Betts, M. M. 1958. The behaviour of adult tits toward other birds and mammals near the nest. *British Birds* 51: 426–429.
Birkhead, M. E. 1981. The social behaviour of the Dunnock *Prunella modularis*. *Ibis* 123: 75–84.
Bleitz, D. 1951. Nest of Pygmy Nuthatches attended by four parents. *Condor* 53: 150–151.
Brackbill, H. 1943. A nesting study of the Wood Thrush. *Wilson Bull*. 55: 73–87.
———. 1944. Juvenile Cardinal helping at a nest. *Wilson Bull*. 56: 50.
———. 1952. A joint nesting of Cardinals and Song Sparrows. *Auk* 69: 302–307.
———. 1958. Titmouse mother's helper. *Baltimore Evening Sun*, June 18.
Bragg, M. B. 1968. Kingbird feeding Baltimore Oriole nestlings. *Auk* 85: 321.
Brosset, A. 1978. Social organization and nest-building in the forest weaver birds of the genus *Malimbus* (Ploceinae). *Ibis* 120: 27–37.
Brown, A. J., II. 1981. Juvenile Eastern Bluebirds participating in nest building. *Sialia* 3: 7.
Brown, C. R. 1977. Starling feeding Purple Martins. *Southwest Naturalist* 21: 557–558.
———. 1983. Mate replacement in Purple Martins: Little evidence for altruism. *Condor* 85: 106–107.
Brown, J. L. 1963. Social organization and behavior of the Mexican Jay. *Condor* 65: 126–153.
———. 1972. Communal feeding of nestlings in the Mexican Jay (*Aphelocoma ultramarina*): Interflock comparisons. *Anim. Behav*. 20: 395–403.

———. 1974. Alternate routes to sociality in jays—with a theory for the evolution of altruism and communal breeding. *Amer. Zool.* 14: 63–80.
———. 1975. Helpers among Arabian Babblers, *Turdoides squamiceps*. *Ibis* 117: 243–244.
———. 1978. Avian communal breeding systems. *Ann. Rev. Ecol. Syst.* 9: 123–155.
———. 1984. The evolution of helping behavior—an ontogenetic and comparative perspective. In *The evolution of adaptive skills: Comparative and ontogenetic approaches*, ed. E. S. Gollin. Hillsdale, N.J.: L. Erlbeum.
Brown, J. L., and R. P. Balda. 1977. The relationship of habitat quality to group size in Hall's Babbler (*Pomatostomus halli*). *Condor* 79: 312–320.
Brown, J. L., and E. R. Brown. 1980. Reciprocal aid-giving in a communal bird. *Z. Tierpsychol.* 53: 313–324.
——— and ———. 1981. Extended family system in a communal bird. *Science* 211: 959–960.
——— and ———. 1984. Parental facilitation: Parent-offspring relations in communally breeding birds. *Behav. Ecol. Sociobiol.* 14: 203–209.
Brown, J. L., D. D. Dow, E. R. Brown, and S. D. Brown. 1978. Effects of helpers on feeding of nestlings in the Grey-crowned Babbler (*Pomatostomus temporalis*). *Behav. Ecol. Sociobiol.* 4: 43–59.
Brown, L. H. 1958. The breeding biology of the Greater Flamingo *Phoenicopterus ruber* at Lake Elmenteita, Kenya Colony. *Ibis* 100: 388–420.
Brown, R. J., and M. N. Brown. 1980. Cooperative breeding in robins of the genus *Eopsaltria*. *Emu* 80: 89.
——— and ———. 1982. Learning behaviour at the nest of the cooperatively breeding Yellow-rumped Thornbill *Acanthiza chrysorrhoa*. *Emu* 82: 111–112.
Bruning, D. F. 1975. Social structure and reproductive behavior in the Greater Rhea. *Living Bird* 13 (for 1974): 251–294.
Carr, T., and C. J. Goin, Jr. 1965. Bluebirds feeding Mockingbird nestlings. *Wilson Bull.* 77: 405–407.
Cayley, N. W. 1949. *The fairy wrens of Australia*. Sydney and London: Angus and Robertson.
Clarke, M. F. 1984a. Co-operative breeding by the Australian Bell Miner *Manorina melanophrys* Latham: A test of kin selection theory. *Behav. Ecol. Sociobiol.* 14: 137–146.
———. 1984b. Interspecific aggression within the genus *Manorina*. *Emu* 84: 113–115.
Clunie, F. 1976. Behaviour and nesting of Fijian White-breasted Wood-Swallows. *Notornis* 23: 61–75.
Collias, N. E., and E. C. Collias. 1964. Evolution of nest building in the weaverbirds (Ploceidae). *Univ. California Publ. Zool.* 73.
——— and ———. 1978a. Cooperative breeding in the White-browed Sparrow-Weaver. *Auk* 95: 472–484.
——— and ———. 1978b. Nest building and nesting behaviour of the Sociable Weaver *Philetairus socius*. *Ibis* 120: 1–15.
——— and ———. 1980. Behavior of the Grey-capped Social Weaver (*Pseudonigrita arnaudi*) in Kenya. *Auk* 97: 213–226.
Conover, M. R., D. E. Miller, and G. L. Hunt, Jr. 1979. Female-female pairs and other unusual reproductive associations in Ring-billed and California gulls. *Auk* 96: 6–9.
Conway, W. G. 1965. Apartment-building and cliff-dwelling parrots. *Anim. Kingdom* (N.Y. Zool. Soc.) 68: 40–46.
Counsilman, J. J. 1977a. Groups of the Grey-crowned Babbler in central southern Queensland. *Babbler* 1: 14–22.
———. 1977b. A comparison of two populations of the Grey-crowned Babbler [part 1]. *Bird Behav.* (formerly called *The Babbler*) 1: 43–82.

Craig, A. 1983. Co-operative breeding in two African starlings, Sturnidae. *Ibis* 125: 114–115.
Craig, J. L. 1979. Habitat variation in the social organization of a communal gallinule, the Pukeko, *Porphyrio porphyrio melanotus*. *Behav. Ecol. Sociobiol.* 5: 331–358.
———. 1980a. Pair and group breeding behaviour of a communal gallinule, the Pukeko, *Porphyrio p. melanotus*. *Anim. Behav.* 28: 593–603.
———. 1980b. Breeding success of a communal gallinule. *Behav. Ecol. Sociobiol.* 6: 289–295.
Craig, J. L., A. M. Stewart, and J. L. Brown. 1982. Subordinates must wait. *Z. Tierpsychol.* 60: 275–280.
Crockett, A. B., and P. L. Hansley. 1978. Coition, nesting, and postfledging behavior of Williamson's Sapsucker in Colorado. *Living Bird* 16 (for 1977): 7–19.
Crossin, R. S. 1967. The breeding biology of the Tufted Jay. *Proc. Western Found. Vertebrate Zool.* 1: 265–299.
Cullen, J. M. 1957. Plumage, age, and mortality in the Arctic Tern. *Bird Study* 4: 197–207. (Abstract in *Ibis* 102: 336. 1960.)
Curry, J. R. 1969. Red-bellied Woodpecker feeds Tufted Titmouse. *Wilson Bull.* 81: 470.
Darwin, C. 1871. *The descent of man and selection in relation to sex*. London: J. Murray.
Davis, D. E. 1940a. Social nesting habits of the Smooth-billed Ani. *Auk* 57: 179–218.
———. 1940b. Social nesting habits of *Guira guira*. *Auk* 57: 472–484.
———. 1942. The phylogeny of social nesting habits in the Crotophaginae. *Quarterly Rev. Biol.* 17: 115–134.
Davis, M. 1952. Captive Raven carries food to non-captive Black Vulture. *Auk* 69: 201.
Davis, M. F. 1978. A helper at a Tufted Titmouse nest. *Auk* 95: 767.
Deck, R. S. 1945. The neighbors' children. *Nature Mag.* 38: 241–242, 272.
De Garis, C. F. 1936. Notes on six nests of the Kentucky Warbler (*Oporornis formosus*). *Auk* 53: 418–428.
Dexter, R. W. 1952. Extra-parental cooperation in the nesting of Chimney Swifts. *Wilson Bull.* 64: 133–139.
———. 1981. Nesting success of Chimney Swifts related to age and the number of adults at the nest, and the subsequent fate of the visitors. *J. Field Ornith.* 52: 228–232.
Douthwaite, R. J. 1978. Breeding biology of the Pied Kingfisher *Ceryle rudis* on Lake Victoria. *J. E. Africa Nat. Hist. Soc. and Natl. Mus.* 31: 1–12.
Dow, D. D. 1978. Reproductive behavior of the Noisy Miner, a communally breeding honey eater. *Living Bird* 16 (for 1977): 163–185.
———. 1979. Agonistic and spacing behaviour of the Noisy Miner *Manorina melanocephala*, a communally breeding honeyeater. *Ibis* 121: 423–436.
Duebbert, H. F. 1968. Two female Mallards incubating on one nest. *Wilson Bull.* 80: 102.
Dyrcz, A. 1977. Nest-helpers in the Alpine Accentor *Prunella collaris*. *Ibis* 119: 215.
Eddinger, C. R. 1967. Feeding helpers among immature white-eyes. *Condor* 69: 530–531.
———. 1970. The white-eye as an interspecific feeding helper. *Condor* 72: 240.
Ellis, H. R. 1952. Nesting behavior of a Purple-throated Fruit-Crow. *Wilson Bull.* 64: 98–100.
Emlen, S. 1978. The evolution of cooperative breeding in birds. In *Behavioural ecology: An evolutionary approach*, eds. J. Krebs and N. Davies. London: Blackwell.
Emlen, S. T., and N. J. Demong. 1984. Bee-eaters of Baharini. *Nat. Hist.* 93: 51–58.
Ervin, S. 1977. Flock size, composition, and behavior in a population of Bushtits (*Psaltriparus minimus*). *Bird-Banding* 48: 97–109.
———. 1978. Bushtit helpers: Accident or altruism? *Bird Behav.* 1: 93–97.
Faaborg, J., T. de Vries, C. B. Patterson, and C. R. Griffin. 1980. Preliminary observations on the occurrence and evolution of polyandry in the Galápagos Hawk (*Buteo galapagoensis*). *Auk* 97: 581–590.

ffrench, R. 1973. *A guide to the birds of Trinidad and Tobago.* Wynnewood, Pa.: Livingston Publishing Co.
Finley, W. L. 1907. *American birds.* New York: C. Scribner's Sons.
Fischer, R. B. 1958. The breeding biology of the Chimney Swift *Chaetura pelagica* (Linnaeus). *New York State Mus. and Sci. Service Bull.* 368.
Fitch, H. S. 1949. Sparrow adopts kingbirds. *Auk* 66: 368–369.
Fjeldså, J. 1981. Biological notes on the Giant Coot *Fulica gigantea. Ibis* 123: 423–437.
Forbush, E. H. 1927. *Birds of Massachusetts and other New England states.* Vol. 2. Boston: Massachusetts Department of Agriculture.
———. 1929. *Birds of Massachusetts and other New England states.* Vol. 3. Boston: Massachusetts Department of Agriculture.
Ford, J. 1963. Breeding behaviour of the Yellow-tailed Thornbill in south-western Australia. *Emu* 63: 185–200.
Forshaw, J. M. 1977. *Parrots of the world.* Neptune, N.J.: T. F. H. Publications.
Fox, W. 1952. Behavioral and evolutionary significance of the abnormal growth of beaks of birds. *Condor* 54: 160–162.
Fraga, R. 1972. Cooperative breeding and a case of successive polyandry in the Bay-winged Cowbird. *Auk* 89: 447–449.
———. 1979. Helpers at the nest in passerines from Buenos Aires Province, Argentina. *Auk* 96: 606–608.
———. 1980. The breeding of Rufous Horneros (*Furnarius rufus*). *Condor* 82: 58–68.
Friedmann, H. 1960. The parasitic weaverbirds. *U.S. Natl. Mus. Bull.* 223.
———. 1968. The evolutionary history of the avian genus *Chrysococcyx. U.S. Natl. Mus. Bull.* 265.
Frith, H. J. 1956. Breeding habits in the family Megapodiidae. *Ibis* 98: 620–640.
Frith, H. J., and S. J. J. F. Davies. 1961. Ecology of the Magpie Goose, *Anseranas semipalmata* Latham (Anatidae). *C.S.I.R.O. Wildlife Research* (Australia) 6: 91–141.
Fry, C. H. 1967. Studies of bee-eaters. *Nigerian Field* 32: 4–17. (Abstract in *Auk* 84: 454–455. 1967.)
———. 1972. The social organisation of bee-eaters (Meropidae) and co-operative breeding in hot-climate birds. *Ibis* 114: 1–14.
———. 1980. The evolutionary biology of kingfishers (Alcedinidae). *Living Bird* 18: 113–160.
Gaston, A. J. 1973. The ecology and behaviour of the Long-tailed Tit. *Ibis* 115: 330–351.
———. 1977. Social behaviour within groups of Jungle Babblers (*Turdoides striatus*). *Anim. Behav.* 25: 828–848.
———. 1978a. Social behaviour of the Yellow-eyed Babbler *Chrysomma sinensis. Ibis* 120: 361–364.
———. 1978b. Ecology of the Common Babbler *Turdoides caudatus. Ibis* 120: 415–432.
———. 1978c. Demography of the Jungle Babbler *Turdoides striatus. J. Anim. Ecol.* 47: 845–870. (Abstract in *Ibis* 121: 392. 1979.)
Gill, F. B. 1971. Ecology and evolution of sympatric Mascarene White-eyes, *Zosterops borbonica* and *Zosterops olivacea. Auk* 88: 35–60.
Gilliard, E. T. 1958. *Living birds of the world.* Garden City, N.Y.: Doubleday and Co.
Goodwin, D. 1947. Breeding behaviour in domestic pigeons four weeks old. *Ibis* 89: 656–658.
———. 1976. *Crows of the world.* Ithaca, N.Y.: Comstock Publishing Associates, Cornell University Press.
Grant, P. R., and N. Grant. 1979. Breeding and feeding of Galápagos Mockingbirds. *Auk* 96: 723–736.

Greenlaw, J. S. 1977. White-throated Sparrow as foster parent of fledgling Dark-eyed Juncos. *Bird-Banding* 48: 170–171.

Greig-Smith, P. W. 1979. Observations of nesting and group behaviour of Seychelles White-eyes *Zosterops modesta*. *Ibis* 121: 344–348.

Grey of Fallodon. 1927. *The charm of birds*. New York: Frederick A. Stokes Co.

Grimes, L. G. 1976. The occurrence of cooperative breeding behaviour in African birds. *Ostrich* 47: 1–15.

———. 1980. Observations of group behaviour and breeding biology of the Yellow-billed Shrike *Corvinella corvina*. *Ibis* 122: 166–192.

Grimes, S. A. 1940. Scrub Jay reminiscences. *Bird-Lore* 42: 431–436.

Grimmer, J. L. 1962. Strange little world of the Hoatzin. *Natl. Geog. Mag.* 122: 391–400.

Gross, A. O. 1949. Nesting of the Mexican Jay in the Santa Rita Mountains, Arizona. *Condor* 51: 241–249.

Haartman, L. von. 1953. Was reizt den Trauerfliegenschnäpper (*Muscicapa hypoleuca*) zu füttern? *Vogelwarte* 16: 157–164.

———. 1956. Territory in the Pied Flycatcher *Muscicapa hypoleuca*. *Ibis* 98: 460–475.

Hales, H. 1896. Peculiar traits of some Scarlet Tanagers. *Auk* 13: 261–263.

Hamilton, G. D. 1952. English Sparrow feeding young Eastern Kingbirds. *Condor* 54: 316.

Hamilton, R. B. 1975. *Comparative behavior of the American Avocet and the Black-necked Stilt* (Recurvirostridae). Amer. Ornith. Union, Ornith. Monogr. 17.

Hardy, J. W. 1961. Studies in behavior and phylogeny of certain New World jays (Garrulinae). *Univ. Kansas Sci. Bull.* 42: 13–149.

———. 1976. Comparative breeding behavior and ecology of the Bushy-crested and Nelson San Blas jays. *Wilson Bull.* 88: 96–120.

Hardy, J. W., T. A. Webber, and R. J. Raitt. 1981. Communal social biology of the Southern San Blas Jay. *Bull. Florida State Mus., Biol. Sci.* 26: 203–263.

Harrison, C. J. O. 1969. Helpers at the nest in Australian passerine birds. *Emu* 69: 30–40.

Hatch, J. J. 1966. Collective territories in Galápagos Mockingbirds, with notes on other behavior. *Wilson Bull.* 78: 198–206.

Hawksley, O., and A. P. McCormack. 1951. Doubly-occupied nests of the eastern Cardinal, *Richmondena cardinalis*. *Auk* 68: 515–516.

Herbert, K. G. S. 1971. Starling feeds young robins. *Wilson Bull.* 83: 316–317.

Hickling, R. A. O. 1959. The burrow-excavation phase in the breeding cycle of the Sand Martin *Riparia riparia*. *Ibis* 101: 497–500.

Hochbaum, H. A. 1960. The brood season. *Nat. Hist.* 69: 54–61.

Hofman, D. E. 1980. Feeding of nestling Cliff Swallows by a House Sparrow. *Canadian Field Natur.* 94: 462.

Hooper, R. G., and M. R. Lennartz. 1983. Roosting behavior of Red-cockaded Woodpecker clans with insufficient cavities. *J. Field Ornith.* 54: 72–76.

Houck, W. J., and J. H. Oliver. 1954. Unusual nesting behavior of the Brown-headed Nuthatch. *Auk* 71: 330–331.

Howard, L. 1952. *Birds as individuals*. London: Collins.

Howe, H. F. 1979. Evolutionary aspects of parental care in the Common Grackle, *Quiscalus quiscula*. *Evolution* 33: 41–51.

Howe, R. W., and R. A. Noske. 1980. Cooperative feeding of fledglings by Crested Shrike-Tits. *Emu* 80: 40.

Hoyt, J. S. Y. 1948. Observations on nesting associates. *Auk* 65: 188–196.

Hudson, W. H. 1920. *Birds of La Plata*. 2 vols. London: J. M. Dent and Sons.

Huels, T. R. 1981. Cooperative breeding in the Golden-breasted Starling *Cosmopsarus regius*. *Ibis* 123: 539–542.
Hutson, H. P. W. 1947. Observations on the nesting of some birds around Delhi. *Ibis* 89: 569–576.
Immelmann, K. 1966. Beobachtungen an Schwalbenstaren. *J. für Ornith.* 107: 37–69.
Ivor, H. R. 1944a. Bird study and semi-captive birds: The Rose-breasted Grosbeak. *Wilson Bull.* 56: 91–104.
———. 1944b. Aye, she was Bonnie. *Nature Mag.* 37: 473–476.
Jackson, J. A. 1974. Gray Rat Snakes versus Red-cockaded Woodpeckers: Predator-prey adaptations. *Auk* 91: 342–347.
Jehl, J. R., Jr. 1968. The breeding biology of Smith's Longspur. *Wilson Bull.* 80: 123–149.
———. 1973. Breeding biology and systematic relationships of the Stilt Sandpiper. *Wilson Bull.* 85: 115–147.
Joste, N. E., W. D. Koenig, R. L. Mumme, and F. A. Pitelka. 1982. Intragroup dynamics of a cooperative breeder: An analysis of reproductive roles in the Acorn Woodpecker. *Behav. Ecol. Sociobiol.* 11: 195–201.
Kale, W. H., II. 1962. A captive Marsh Wren helper. *Oriole*, June (separate not paged).
Kemp, A. C. 1979. A review of the hornbills: Biology and radiation. *Living Bird* 17 (for 1978): 105–136.
Kemp, A. C., and M. I. Kemp. 1980. The biology of the Southern Ground Hornbill *Bucorvus leadbeateri* (Vigors) (Aves: Bucerotidae). *Ann. Transvaal Mus.* 32: 65–100.
Kendeigh, S. C. 1945. Nesting behavior of wood warblers. *Wilson Bull.* 57: 145–164.
Kepler, A. K. 1977. *Comparative study of todies (Todidae): With emphasis on the Puerto Rican Tody,* Todus mexicanus. Publ. Nuttall Ornith. Club 16. Cambridge, Mass.
Keppie, D. M. 1977. Inter-brood movements of juvenile Spruce Grouse. *Wilson Bull.* 89: 67–72.
Kilham, H. 1956. Breeding and other habits of Casqued Hornbills (*Bycanistes subcylindricus*). *Smithsonian Misc. Coll.* 131 (9): 1–45.
———. 1977. Altruism in nesting Yellow-bellied Sapsucker. *Auk* 94: 613–614.
Kiltie, R. A., and J. W. Fitzpatrick. 1984. Reproduction and social organization of the Black-capped Donacobius (*Donacobius atricapillus*) in southeastern Perú. *Auk* 101: 804–811.
King, B. R. 1980. Social organization and behaviour of the Grey-crowned Babbler *Pomatostomus temporalis*. *Emu* 80: 59–76.
Kinnaird, M. F., and P. R. Grant. 1982. Cooperative breeding by the Galápagos Mockingbird, *Nesomimus parvulus*. *Behav. Ecol. Sociobiol.* 10: 65–73.
Koenig, W. D. 1981a. Reproductive success, group size, and the evolution of cooperative breeding in the Acorn Woodpecker. *Amer. Naturalist* 117: 421–443.
———. 1981b. Space competition in the Acorn Woodpecker: Power struggles in a cooperative breeder. *Anim. Behav.* 29: 396–409.
Koenig, W. D., R. L. Mumme, and F. A. Pitelka. 1983. Female roles in cooperatively breeding Acorn Woodpeckers. In *Social behavior of female vertebrates*, ed. S. K. Wasser. New York: Academic Press.
Koenig, W. D., and F. A. Pitelka. 1979. Relatedness and inbreeding avoidance in the communally nesting Acorn Woodpecker. *Science* 206: 1103–1105.
Köster, F. 1971. Zum Nistverhalten des Ani, *Crotophaga ani*. *Bonn. Zool. Beitr.* 22: 4–27.
Krekorian, C. O. 1978. Alloparental care in the Purple Gallinule. *Condor* 80: 382–390.
Kropotkin, P. 1902. *Mutual aid: A factor in evolution*. London: William Heinemann.
Lack, D. 1953. *The life of the Robin*. London: Pelican Books.
———. 1968. *Ecological adaptations for breeding in birds*. London: Methuen and Co.

Lack, D., and E. Lack. 1958. The nesting of the Long-tailed Tit. *Bird Study* 5: 1–19.
Lashley, K. S. 1938. Experimental analysis of instinctive behavior. *Psychol. Rev.* 45: 445–471.
Laskey, A. R. 1939. A study of nesting Eastern Bluebirds. *Bird-Banding* 10: 23–32.
———. 1948. Some nesting data on the Carolina Wren at Nashville, Tennessee. *Bird-Banding* 19: 101–121.
———. 1957. Some Tufted Titmouse life history. *Bird-Banding* 28: 135–145.
Lawrence, L. de K. 1947. Five days with a pair of nesting Canada Jays. *Canadian Field Natur.* 61: 1–11.
Lawton, M. F., and C. F. Guindon. 1981. Flock composition, breeding success, and learning in the Brown Jay. *Condor* 83: 27–33.
Lay, D. W., and D. N. Russell. 1970. Notes on the Red-cockaded Woodpecker in Texas. *Auk* 87: 781–786.
Lefebvre, E. A. 1977. Laysan Albatross breeding behavior. *Auk* 94: 270–274.
Lemmons, P. 1956. Cardinal feeds fishes. *Nature Mag.* 49: 536.
Lennartz, M. R., and R. F. Harlow. 1979. The role of parent and helper Red-cockaded Woodpeckers at the nest. *Wilson Bull.* 91: 331–335.
Lewis, D. M. 1982a. Cooperative breeding in a population of White-browed Sparrow Weavers. *Ibis* 124: 511–522.
———. 1982b. Dispersal in a population of White-browed Sparrow Weavers. *Condor* 84: 306–312.
Ligon, J. D. 1970. Behavior and breeding biology of the Red-cockaded Woodpecker. *Auk* 87: 255–278.
Ligon, J. D., and S. L. Husar. 1974. Notes on the behavioral ecology of Couch's Mexican Jay. *Auk* 91: 841–843.
Ligon, J. D., and S. H. Ligon. 1979. The communal social system of the Green Woodhoopoe in Kenya. *Living Bird* (for 1978) 17: 159–197.
Lind, E. A. 1960. Zur Ethologie und Ökologie der Mehlschwalbe, *Delichon* u. *urbica* (L.). *Ann. Zool. Soc. "Vanamo"* 21: 1–123.
———. 1964. Nistzeitliche Geselligkeit des Mehlschwalbe *Delichon* u. *urbica* (L.). *Ann. Zool. Fenn.* 1: 7–43.
Lindgren, F. 1975. [Observations on the Siberian Jay *Perisoreus infaustus*—its breeding biology in particular.] *Fauna Flora Upps.* 70: 198–210. (Swedish.) (Abstract in *Ibis* 118: 283. 1976.)
Logan, S. 1951. Cardinal, *Richmondena cardinalis*, assists in feeding robins. *Auk* 68: 516–517.
Lonsdale, W. S. 1935. Blue Tits feeding young Robins. *British Birds* 29: 113–114.
Maciula, S. J. 1960. Worm-eating Warbler "adopts" Ovenbird nestlings. *Auk* 77: 220.
Maclean, G. L. 1973. The Sociable Weaver, parts 1–5. *Ostrich* 44: 176–261.
MacRoberts, M. H., and B. R. MacRoberts. 1976. *Social organization and behavior of the Acorn Woodpecker in central coastal California*. Amer. Ornith. Union, Ornith. Monogr. 21.
Mader, W. J. 1975. Extra adults at Harris' Hawk nests. *Condor* 77: 482–485.
———. 1976. Biology of the Harris' Hawk in southern Arizona. *Living Bird* 14 (for 1975): 59–85.
———. 1979. Breeding behavior of a polyandrous trio of Harris' Hawks in southern Arizona. *Auk* 96: 776–788.
Marchant, S. 1960. The breeding of some S.W. Ecuadorian birds. *Ibis* 102: 349–382.
———. 1984. Notes on the breeding of Varied Sittellas *Daphoenositta chrysoptera*. *Corella* 8: 11–15. (Abstract in *Auk* 102, 1, suppl.: 20A–21A. 1985.)
Marcström, V., and F. Sundgren. 1977. On the reproduction of the European Woodcock. *Viltrevy* (Swedish Wildl.) 10: 27–40. (Abstract in *Auk* 95, 3, suppl.: 12C. 1978.)
Martin, J. 1968. Karoo Prinias feeding Layard's Tit-Babbler. *Ostrich* 39: 263.

Mathews, G. M. 1924. *The birds of Australia*. Vol. 11. London: Witherby and Co.
Maxson, S. J. 1978. Evidence of brood adoption by Ruffed Grouse. *Wilson Bull.* 90: 132–133.
Messenger, W. H. 1949. Goose baby-sitter. *Nature Mag.* 42: 79.
Mills, E. A. 1931. *Bird memories of the Rockies*. Boston and New York: Houghton Mifflin Co.
Milne, H. 1969. Eider biology. *Ibis* 111: 278.
Miskell, J. 1977. Co-operative feeding of young at the nest by Fischer's Starling *Spreo fischeri*. *Scopus* 1: 87–88.
Moermond, T. C. 1981. Cooperative feeding, defense of young, and flocking in the Black-faced Grosbeak. *Condor* 83: 82–83.
Mohr, H. 1958. Ein Fall von Polygamie bei der Rauchschwalbe (*Hirundo rustica*). *Orn. Mitteil.* 10: 7–9.
Moore, R. T. 1938. Discovery of the nest and eggs of the Tufted Jay. *Condor* 40: 233–241.
Moreau, R. E. 1936. The breeding biology of certain East African hornbills (Bucerotidae). *J. E. Africa and Uganda Nat. Hist. Soc.* 13: 1–28.
Moreau, R. E., and W. M. Moreau. 1937. Biological and other notes on some East African birds. *Ibis*, ser. 14, 1: 152–174.
———. 1940. Hornbill studies. *Ibis*, ser. 14, 4: 639–656.
Morrison, M. L., and R. D. Slack. 1977. Flocking and foraging behavior of Brown Jays in northeastern Mexico. *Wilson Bull.* 89: 171–173.
Mountfort, G. 1957. *The Hawfinch*. London: Collins.
Munro, J., and J. Bédard. 1977a. Gull predation and crèching behavior in the Common Eider. *J. Anim. Ecol.* 46: 799–810. (Abstract in *Ibis* 121: 399. 1979.)
——— and ———. 1977b. Crèche formation in the Common Eider. *Auk* 94: 759–771.
Murphy, R. C. 1936. *Oceanic birds of South America*. New York: American Museum of Natural History.
Murton, R. K., and A. J. Isaacson. 1962. The functional basis of some behaviour in the Woodpigeon *Columba palumbus*. *Ibis* 104: 503–521.
Myers, G. R., and D. W. Waller. 1977. Helpers at the nest in Barn Swallows. *Auk* 94: 596.
Narosky, S., R. Fraga, and M. de la Peña. 1983. *Nidificación de las aves argentinas (Dendrocolaptidae y Furnariidae)*. Buenos Aires: Asociación Ornitológica del Plata.
Neff, J. A. 1945. Foster parentage of a Mourning Dove in the wild. *Condor* 47: 39–40.
Nice, M. M. 1943. Studies in the life history of the Song Sparrow. 2. *Trans. Linn. Soc. New York* 6: 1–328.
———. 1962. Development of behavior in precocial birds. *Trans. Linn. Soc. New York* 8: i–xii, 1–211.
Nicholson, E. M. 1930. Field notes on Greenland birds. *Ibis*, ser. 12, 6: 280–313, 395–428.
Nolan, V., Jr. 1965. A male Cardinal helper at a nest of Yellow-breasted Chats. *Wilson Bull.* 77: 196.
———. 1978. *The ecology and behavior of the Prairie Warbler* Dendroica discolor. Amer. Ornith. Union, Ornith. Monogr. 26.
Nolan, V., Jr., and R. Schneider. 1962. A Catbird helper at a House Wren nest. *Wilson Bull.* 74: 183–184.
Norris, R. A. 1958. Comparative biosystematics and life history of the nuthatches *Sitta pygmaea* and *Sitta pusilla*. *Univ. California Publ. Zool.* 56: 119–300.
Noske, R. A. 1980a. Co-operative breeding and plumage variation in the Orange-winged (Varied) Sittella. *Corella* 4: 45–53.
———. 1980b. Cooperative breeding by treecreepers. *Emu* 80: 35–36.
———. 1982. Comparative behaviour and ecology of some Australian bark-foraging birds. Ph.D. thesis, University of New England, Armidale, N.S.W.
———. 198?. The private lives of treecreepers. *Australian Nat. Hist.* 20: 419–424.

———. 1983. Communal behaviour of Brown-headed Honeyeaters. *Emu* 83: 38–41.
Orians, G. H., L. Erckmann, and J. C. Schultz. 1977. Nesting and other habits of the Bolivian Blackbird (*Oreospar bolivianus*). *Condor* 79: 250–256.
Orians, G. H., C. E. Orians, and K. J. Orians. 1977. Helpers at the nest in some Argentine blackbirds. In *Evolutionary ecology*, eds. B. Stonehouse and C. Perrins. London and New York: Macmillan.
Parry, V. 1971. *Kookaburras*. Melbourne: Lansdowne Press.
———. 1973. The auxiliary social system and its effect on territory and breeding in Kookaburras. *Emu* 73: 81–100.
Perry, R. 1946. *Lundy: Isle of Puffins*. London: Lindsay Drummond.
Pinkowski, B. C. 1975. Yearling male Eastern Bluebird assists parents in feeding young. *Auk* 92: 801–802.
———. 1978. Two successive male Eastern Bluebirds tending the same nest. *Auk* 95: 606–608.
Poole, A. 1982. Breeding Ospreys feed fledglings that are not their own. *Auk* 99: 781–784.
Poonswad, P., A. Tsuji, and C. Nagarmpongsai. 1983. A study of the breeding biology of hornbills (Bucerotidae) in Thailand. *Proc. Delacour-IFCB Symposium on Breeding Birds in Captivity*: 239–265.
Powell, H. 1946. Nuthatch feeding nestling Starlings. *British Birds* 39: 316. (Abstract in *Ibis* 89: 152. 1947.)
Prescott, K. W. 1965. Studies in the life history of the Scarlet Tanager, *Piranga olivacea*. Trenton: New Jersey State Museums, Investigations 2.
———. 1967. Unusual activities of a House Sparrow and a Blue Jay at a Tufted Titmouse nest. *Wilson Bull.* 79: 346–347.
———. 1971. Unusual activity of Starlings at a Yellow-shafted Flicker nest. *Wilson Bull.* 83: 195–196.
Price, T., S. Millington, and P. Grant. 1983. Helping at the nest of Darwin's finches as misdirected parental care. *Auk* 100: 192–194.
Pullman, J. O. 1970. A Tufted Titmouse nest attended by Carolina Chickadees. *Chat* 34: 22. (Abstract in *Auk* 87: 834. 1970.)
Raitt, R. J., and J. W. Hardy. 1976. Behavioral ecology of the Yucatán Jay. *Wilson Bull.* 88: 529–554.
——— and ———. 1979. Social behavior, habitat, and food of the Beechey Jay. *Wilson Bull.* 91: 1–15.
Raitt, R. J., S. R. Winterstein, and J. W. Hardy. 1984. Structure and dynamics of communal groups in the Beechey Jay. *Wilson Bull.* 96: 206–227.
Rakestraw, J. L., and J. L. Baker. 1981. Dusky Seaside Sparrow feeds Red-winged Blackbird fledglings. *Wilson Bull.* 93: 540.
Rand, A. L. 1953. Factors affecting feeding rates of anis. *Auk* 70: 26–30.
Raney, E. C. 1939. Robin and Mourning Dove use same nest. *Auk* 56: 337–338.
Rea, G. 1945. Black and White Warbler feeding young of Worm-eating Warbler. *Wilson Bull.* 57: 262.
Reyer, H.-U. 1980a. Flexible helper structure as an ecological adaptation in the Pied Kingfisher (*Ceryle rudis rudis* L.). *Behav. Ecol. Sociobiol.* 6: 219–227.
———. 1980b. Sexual dimorphism and co-operative breeding in the Striped Kingfisher. *Ostrich* 51: 117–118.
Reynolds, J. F. 1974. Cooperative breeding in Red-and-Yellow Barbets. *Bull. E. Africa Nat. Hist. Soc.* 1974: 144–145. (Abstract in *Ibis* 118: 142. 1976.)
Rice, D. W., and K. W. Kenyon. 1962. Breeding cycles and behavior of Laysan and Black-footed albatrosses. *Auk* 79: 517–567.

Rice, O. O. 1969. Record of female Cardinals sharing nest. *Wilson Bull.* 81: 216.
Richmond, S. M. 1953. The attraction of Purple Martins to an urban location in western Oregon. *Condor* 55: 225–249.
Ricklefs, R. E. 1975. The evolution of cooperative breeding in birds. *Ibis* 117: 531–534.
———. 1980. 'Watch-dog' behaviour observed at the nest of a cooperative breeding bird, the Rufous-margined Flycatcher *Myiozetetes cayanensis. Ibis* 122: 116–118.
Ridpath, M. G. 1964. The Tasmanian Native Hen. *Australian Nat. Hist.* 14: 346–350.
Riney, T. 1951. Relationships between birds and deer. *Condor* 53: 178–185.
Robinson, G. G. 1962. Winter Wren feeds Townsend Solitaire young. *Condor* 64: 240.
Rowley, I. 1957. The cooperative feeding of young by Superb Blue Wrens. *Emu* 57: 356–357.
———. 1965. The life history of the Superb Blue Wren *Malurus cyaneus. Emu* 64: 251–297.
———. 1975. *Bird life.* Sydney and London: Collins.
———. 1976. Co-operative breeding in Australian birds. *Proc. 16th Internatl. Ornith. Congr.*: 657–666. Canberra City: Australian Academy of Sciences.
———. 1978. Communal activities among White-winged Choughs *Corcorax melanorhamphus. Ibis* 120: 178–197.
———. 1981a. The communal way of life of the Splendid Wren *Malurus splendens. Z. Tierpsychol.* 55: 228–267.
———. 1981b. A relic population of Blue-breasted Wrens *Malurus pulcherrimus* in the central wheatbelt. *Western Australian Naturalist* 15: 1–8.
Royall, W. C., and R. E. Pillmore. 1968. House Wren feeds Red-shafted Flicker nestlings. *Murrelet* 49: 4–6.
Russell, W. C. 1947. Mountain Chickadees feeding young Williamson's Sapsuckers. *Condor* 49: 83.
Ryder, J. P., and P. L. Somppi. 1979. Female-female pairing in Ring-billed Gulls. *Auk* 96: 1–5.
Sappington, J. E. 1977. Breeding biology of House Sparrows in north Mississippi. *Wilson Bull.* 89: 300–309.
Sauer, E. G., and E. M. Sauer. 1966. The behavior and ecology of the South African Ostrich. *Living Bird* 5: 45–75.
Schifter, H. 1972. Familie Bartvögel. In *Grzimeks Tierleben* 9: 63–75. Munich: Kindler Verlag.
Selander, R. K. 1964. Speciation in wrens of the genus *Campylorhynchus. Univ. California Publ. Zool.* 74: 1–259.
Selous, E. 1927. *Realities of bird life: Being extracts from the diaries of a life-loving naturalist.* London: Constable and Co.
Senar, J. C. 1984. Allofeeding in Eurasian Siskins (*Carduelis spinus*). *Condor* 86: 213–214.
Sherley, G. H. 1985. New Zealand wrens. In *The Encyclopaedia of Birds,* eds. C. M. Perrins and L. A. Middleton. London: George Allen and Unwin.
Sherman, A. R. 1952. *Birds of an Iowa dooryard.* Boston: Christopher Publishing House.
Sherrill, D. M., and V. M. Case. 1980. Winter home ranges of four clans of Red-cockaded Woodpeckers in the Carolina Sandhills. *Wilson Bull.* 92: 369–375.
Short, L. L., Jr. 1964. Extra helpers feeding young of the Blue-winged and Golden-winged warblers. *Auk* 81: 428–430.
———. 1970. Notes on the habits of some Argentine and Peruvian woodpeckers (Aves, Picidae). *Amer. Mus. Novitates* 2413: 1–37.
———. 1974. Habits of three endemic West Indian woodpeckers (Aves, Picidae). *Amer. Mus. Novitates* 2549: 1–44.
Short, L. L., and J. F. M. Horne. 1980. Ground barbets of East Africa. *Living Bird* 18: 179–186.
——— and ———. 1984. Behavioural notes on the White-eared Barbet *Stactolaema leucotis* in Kenya. *Bull. Brit. Ornith. Club* 104: 47–53.

Shugart, G. W. 1980. Frequency and distribution of polygyny in Great Lakes Herring Gulls in 1978. *Condor* 82: 426–429.
Shy, M. M. 1982. Interspecific feeding among birds: A review. *J. Field Ornith.* 53: 370–393.
Sibley, C. G., and J. E. Ahlquist. 1973. The relationship of the Hoatzin. *Auk* 90: 1–13.
Sick, H. 1959. Notes on the biology of two Brazilian swifts, *Chaetura andrei* and *Chaetura cinereiventris*. *Auk* 76: 471–477.
Siegfried, W. R., and P. G. H. Frost. 1975. Continuous breeding and associated behaviour in the Moorhen *Gallinula chloropus*. *Ibis* 117: 102–109.
Skead, C. J., and G. A. Ranger. 1958. A contribution to the biology of the Cape Province white-eyes (*Zosterops*). *Ibis* 100: 319–333.
Skutch, A. F. 1935. Helpers at the nest. *Auk* 52: 257–273.
———. 1953a. The White-throated Magpie-Jay. *Wilson Bull.* 65: 68–74.
———. 1953b. Life history of the Southern House-Wren. *Condor* 55: 121–149.
———. 1954. *Life histories of Central American birds. 1.* Pacific Coast Avifauna 31. Berkeley, Calif.: Cooper Ornithological Society.
———. 1958. Roosting and nesting of araçari toucans. *Condor* 60: 201–219.
———. 1959. Life history of the Groove-billed Ani. *Auk* 76: 281–317.
———. 1960. *Life histories of Central American birds. 2.* Pacific Coast Avifauna 34. Berkeley, Calif.: Cooper Ornithological Society.
———. 1961. Helpers among birds. *Condor* 63: 198–226.
———. 1962. Life histories of honeycreepers. *Condor* 64: 92–125.
———. 1966. Life history notes on three tropical American cuckoos. *Wilson Bull.* 78: 139–165.
———. 1967. Adaptive limitation of the reproductive rate of birds. *Ibis* 109: 579–599.
———. 1968. The nesting of some Venezuelan birds. *Condor* 70: 66–82.
———. 1969a. A study of the Rufous-fronted Thornbird and associated birds. *Wilson Bull.* 81: 5–43, 123–139.
———. 1969b. *Life histories of Central American birds. 3.* Pacific Coast Avifauna 35. Berkeley, Calif.: Cooper Ornithological Society.
———. 1972. *Studies of tropical American birds.* Publ. Nuttall Ornith. Club 10. Cambridge, Mass.
———. 1976. *Parent birds and their young.* Austin: University of Texas Press.
———. 1980. *A naturalist on a tropical farm.* Berkeley and Los Angeles: University of California Press.
———. 1981. *New studies of tropical American birds.* Publ. Nuttall Ornith. Club 19. Cambridge, Mass.
———. 1983a. *Birds of tropical America.* Austin: University of Texas Press.
———. 1983b. *Nature through tropical windows.* Berkeley and Los Angeles: University of California Press.
Smith, S. M. 1971. The relationship of grazing cattle to foraging rates in anis. *Auk* 88: 876–880.
———. 1972. Roosting aggregations of Bushtits in response to cold temperatures. *Condor* 74: 478–479.
Snow, B. K. 1963. The behaviour of the Shag. *British Birds* 56: 77–186.
Snow, D. W. 1958. The breeding of the Blackbird *Turdus merula* at Oxford. *Ibis* 100: 1–30.
———. 1971. Observations on the Purple-throated Fruit-Crow in Guyana. *Living Bird* 10: 5–17.
———. 1982. *The cotingas.* London: British Museum (Natural History); Oxford: Oxford University Press.
Snow, D. W., and C. T. Collins. 1962. Social breeding behavior of the Mexican Tanager. *Condor* 64: 161.

Southern, J. 1952. Spotted Flycatchers feeding nestling Blackbirds. *British Birds* 45: 366.
Southern, W. E. 1959. Foster-feeding and polygamy in the Purple Martin. *Wilson Bull.* 71: 96.
———. 1968. Further observations on foster-feeding by Purple Martins. *Wilson Bull.* 80: 234–235.
Stacey, P. B. 1979a. Habitat saturation and communal breeding in the Acorn Woodpecker. *Anim. Behav.* 27: 1153–1166.
———. 1979b. Kinship, promiscuity, and communal breeding in the Acorn Woodpecker. *Behav. Ecol. Sociobiol.* 6: 53–66.
———. 1981. Foraging behavior of the Acorn Woodpecker in Belize, Central America. *Condor* 83: 336–339.
Stacey, P. B., and W. D. Koenig. 1984. Cooperative breeding in the Acorn Woodpecker. *Scientific Amer.* 251: 114–121.
Stonehouse, B. 1953. The Emperor Penguin *Aptenodytes fosteri* Gray. 1: Breeding behaviour and development. *Falkland Islands Dependencies Survey, Sci. Reports* 6: 1–33. London: H. M. Stationery Office.
———. 1960. The King Penguin *Aptenodytes patagonica* of South Georgia. 1: Breeding behaviour and development. *Falkland Islands Dependencies Survey, Sci. Reports* 23: 1–81. London: H. M. Stationery Office.
———. 1962. Tropic birds (genus *Phaethon*) of Ascension Island. *Ibis* 103b: 124–161.
Strosnider, R. 1960. Polygyny and other notes on the Red-winged Blackbird. *Wilson Bull.* 72: 200.
Stutterheim, C. J. 1982. Breeding biology of the Redbilled Oxpecker in the Kruger National Park. *Ostrich* 53: 79–90.
Sutton, G. M., and D. F. Parmelee. 1954. The nesting of the Greenland Wheatear on Baffin Island. *Condor* 56: 295–306.
Swainson, G. W. 1970. Co-operative rearing in the Bell Miner. *Emu* 70: 183–188.
Tamiya, Y., and M. Aoyanagi. 1982. The significance of reoccupation by non-breeding birds in the Adelie Penguin *Pygoscelis adeliae* during their incubation, guard and crèche periods. *J. Yamashina Inst. Ornith.* 14: 35–44.
Tarbell, A. T. 1983. A yearling helper with a Tufted Titmouse brood. *J. Field Ornith.* 54: 89.
Tarboton, W. 1976. Martial Eagles: An unusual breeding episode. *Bokmakierie* 28: 29–32. (Abstract in *Auk* 94, 4, suppl.: 12D. 1977.)
Tarboton, W. R. 1981. Cooperative breeding and group territoriality in the Black Tit. *Ostrich* 52: 216–225.
Terres, J. K. 1980. *Encyclopedia of North American birds.* New York: Alfred A. Knopf.
Thomas, B. T. 1979. Behavior and breeding of the White-bearded Flycatcher (*Conopias inornata*). *Auk* 96: 767–775.
———. 1983. The Plain-fronted Thornbird: Nest construction, material choice, and nest defense behavior. *Wilson Bull.* 95: 106–117.
Thomson, A. L., ed. 1964. *A new dictionary of birds.* London: Nelson.
Trail, P. W., S. D. Strahl, and J. L. Brown. 1981. Infanticide in relation to individual and flock histories in a communally breeding bird, the Mexican Jay (*Aphelocoma ultramarina*). *Amer. Naturalist* 118: 72–82.
Troetschler, R. G. 1976. Acorn Woodpecker breeding strategy as affected by Starling nest-hole competition. *Condor* 78: 151–165.
Tuck, L. M. 1960. *The murres: Their distribution, populations and biology.* Canadian Wildlife Series 1. Ottawa.
———. 1972. *The snipes: A study of the genus* Capella. Canadian Wildlife Series 5. Ottawa.

Van Riper, C., III. 1980. Observations on the breeding of the Palila *Psittirostra bailleui* of Hawaii. *Ibis* 122: 462–475.
Van Someren, V. G. L. 1956. Days with birds: Studies of some East African species. *Fieldiana: Zoology* 38. Chicago: Chicago Natural History Museum.
Van Velzen, W. T. 1960. Starlings fed by Purple Martin. *Auk* 77: 477.
Vaughan, R. 1980. *Plovers*. Laverham, Suffolk: Terence Dalton Ltd.
Vehrencamp, S. L. 1977. Relative fecundity and parental effort in communally nesting anis, *Crotophaga sulcirostris*. *Science* 197: 403–405.
———. 1978. The adaptive significance of communal nesting in Groove-billed Anis (*Crotophaga sulcirostris*). *Behav. Ecol. Sociobiol.* 5: 1–33.
Verbeek, N. A. M., and R. W. Butler. 1981. Cooperative breeding of the Northwestern Crow *Corvus caurinus* in British Columbia. *Ibis* 123: 183–189.
Vernon, C. J. 1976. Communal feeding of nestlings by the Arrowmarked Babbler. *Ostrich* 47: 134–136. (Abstract in *Ibis* 121: 400. 1979.)
Vernon, F. J., and C. J. Vernon. 1978. Notes on the social behavior of the Duskyfaced Warbler. *Ostrich* 49: 92–93.
Wagner, H. O. 1959. Beitrag zum Verhalten des Weissohrkolibris (*Hylocharis leucotis* Vieill.). *Zool. Jahrb.* 86: 253–302.
Walters, J. 1959. [Observations on two broods of Kentish Plovers, *Charadrius alexandrinus*, on Texel.] *Ardea* 47: 48–67. (Dutch.) (Abstract in *Ibis* 102: 338. 1960.)
Walters, J., and B. F. Walters. 1980. Co-operative breeding by Southern Lapwings *Vanellus chilensis*. *Ibis* 122: 505–509.
Warham, J. 1954. The behaviour of the Splendid Blue Wren. *Emu* 54: 135–140.
———. 1960. Some aspects of breeding behaviour in the Short-tailed Shearwater. *Emu* 60: 75–87.
Weatherhead, P. J. 1979. Behavioral implications of the defense of a shoveler brood by Common Eiders. *Condor* 81: 427.
Wegner, W. A. 1976. Extra-parental assistance by male American Kestrel. *Wilson Bull.* 88: 670.
West-Eberhard, M. J. 1975. The evolution of social behavior by kin-selection. *Quarterly Rev. Biol.* 50: 1–33.
W[estwood], R. W. 1946. Contents noted. *Nature Mag.* 39: 399.
Wiley, J. W. 1975. Three adult Red-tailed Hawks tending a nest. *Condor* 77: 480–482.
Wilkinson, R. 1982. Social organization and communal breeding in the Chestnut-bellied Starling (*Spreo pulcher*). *Anim. Behav.* 30: 1118–1128.
Williams, L. 1942. Interrelations in a nesting group of four species of birds. *Wilson Bull.* 54: 238–249.
Willis, E. 1961. A study of nesting ant-tanagers in British Honduras. *Condor* 63: 479–503.
Witherby, H. F., F. C. R. Jourdain, N. F. Ticehurst, and B. W. Tucker. 1938. *The handbook of British birds*. Vol. 2. London: H. F. and G. Witherby.
Woolfenden, G. E. 1974. Nesting and survival in a population of Florida Scrub Jays. *Living Bird* 12 (for 1973): 25–49.
———. 1975. Florida Scrub Jay helpers at the nest. *Auk* 92: 1–15.
———. 1976. Co-operative breeding in American birds. *Proc. 16th Internatl. Ornith. Congr.*: 674–684. Canberra City: Australian Academy of Sciences.
———. 1978. Growth and survival of Florida Scrub Jays. *Wilson Bull.* 90: 1–18.
Woolfenden, G. E., and J. H. Fitzpatrick. 1977. Dominance in the Florida Scrub Jay. *Condor* 79: 1–12.
——— and ———. 1978. The inheritance of territory in group-breeding birds. *BioScience* 28: 104–107.

Wynne-Edwards, V. C. 1952. Zoology of the Baird expedition (1950). 1: The birds observed in central and southeast Baffin Island. *Auk* 69: 353–391.
Young, E. C. 1963. The breeding behaviour of the South Polar Skua *Catharacta maccormicki*. *Ibis* 105: 203–233.
Young, H. 1955. Breeding behavior and nesting of the eastern Robin. *Amer. Midl. Naturalist* 53: 329–352.
Zahavi, A. 1974. Communal nesting by the Arabian Babbler. *Ibis* 116: 84–87.
———. 1976. Co-operative nesting in Eurasian birds. *Proc. 16th Internatl. Ornith. Congr.*: 685–693. Canberra City: Australian Academy of Sciences.
Zicus, M. C. 1981. Canada Goose brood behavior and survival estimates at Crex Meadows, Wisconsin. *Wilson Bull.* 93: 207–217.

INDEX

Illustrations are indicated by **boldfaced** numerals.